SpringerBriefs in Statistics

AF382535

More information about this series at http://www.springer.com/series/8921

David Fletcher

Model Averaging

 Springer

David Fletcher
Department of Mathematics and Statistics
University of Otago
Dunedin, New Zealand

ISSN 2191-544X ISSN 2191-5458 (electronic)
SpringerBriefs in Statistics
ISBN 978-3-662-58540-5 ISBN 978-3-662-58541-2 (eBook)
https://doi.org/10.1007/978-3-662-58541-2

Library of Congress Control Number: 2018964926

© The Author(s), under exclusive licence to Springer-Verlag GmbH, DE, part of Springer Nature 2018
This work is subject to copyright. All rights are reserved by the Publisher, whether the whole or part of the material is concerned, specifically the rights of translation, reprinting, reuse of illustrations, recitation, broadcasting, reproduction on microfilms or in any other physical way, and transmission or information storage and retrieval, electronic adaptation, computer software, or by similar or dissimilar methodology now known or hereafter developed.
The use of general descriptive names, registered names, trademarks, service marks, etc. in this publication does not imply, even in the absence of a specific statement, that such names are exempt from the relevant protective laws and regulations and therefore free for general use.
The publisher, the authors and the editors are safe to assume that the advice and information in this book are believed to be true and accurate at the date of publication. Neither the publisher nor the authors or the editors give a warranty, express or implied, with respect to the material contained herein or for any errors or omissions that may have been made. The publisher remains neutral with regard to jurisdictional claims in published maps and institutional affiliations.

This Springer imprint is published by the registered company Springer-Verlag GmbH, DE part of Springer Nature
The registered company address is: Heidelberger Platz 3, 14197 Berlin, Germany

To Sonya, for being there...

Preface

In writing this book, my aim has been to provide a succinct, accessible account of model averaging that will be useful to applied statisticians and scientists. I have emphasised the links between methods developed in statistics, econometrics and machine learning, as well as the interplay between the frequentist and Bayesian approaches. I have assumed that the reader is familiar with basic statistical theory and modelling, including probability, likelihood and generalised linear models.

The references should help the reader follow up on topics I have not covered in detail. The number of papers written on model averaging is far greater than I had expected when starting this book, and I apologise in advance if I have overlooked any important articles. I have deliberately chosen small examples to illustrate the different methods, in order to facilitate the discussion of key concepts. Many applications of model averaging will be in more complex settings, but translation of the ideas to those settings will often be clear.

My thoughts on model averaging have benefited greatly from discussions with Richard Barker, Peter Dillingham, Murray Efford, Chuen Yen Hong, Michel de Lange, Matt Parry, Daniel Turek and Jimmy Zeng. I am indebted to Deborah Shaw for her diligent work on formatting the references and to the referees for their helpful comments on an early draft. I am also very grateful to Eva Hiripi of Springer for her editorial help. My colleagues at the University of Otago have provided a most supportive environment within which to work on this book; I am very grateful for their kindness and generosity of spirit.

I owe so much to my wife Sonya, without whom the past 21 years would have been very different. The love and support of our children, Anika, Nils, Kiersten Anna, Thomi, Ana and Daniel, has been priceless, as has that of my sister Anne, her family, and all of our Swiss whanau, especially Elfriede Hamel.

Dunedin, New Zealand
November 2018

David Fletcher

Contents

Chapter 1
Why Model Averaging?

Abstract Model averaging is a means of allowing for model uncertainty in estimation which can provide better estimates and more reliable confidence intervals than model selection. We illustrate its use via examples involving real data, discuss when it is likely to be useful, and compare the frequentist and Bayesian approaches to model averaging.

1.1 Country Fairs and the Size of the Universe

In 1907, Francis Galton, eminent statistician and half-cousin to Charles Darwin, published a paper in Nature [80], the abstract of which begins with the statement

> In these democratic days, any investigation into the trustworthiness and peculiarities of popular judgements is of interest.[1]

He was reporting on the fact that many of the visitors to the recent West of England Fat Stock and Poultry Exhibition had entered a competition to guess the weight of an ox. The mean of the 787 guesses was found to be exactly the same as the true value of 1197 lb.[2] The guesses were made by a mix of farmers, butchers and the general public. Galton suggested that the mixture of abilities to make such a guess would be similar to the mixture of abilities to judge political candidates in an election, a point that prompted him to give his paper the title *Vox Populi*. The fact that the mean was identical to the true figure suggested to him that allowing the whole adult population to vote in an election might have something going for it.

This is a simple example of model averaging, with each person using a "model" to come up with an estimate of the weight. Instead of using a simple mean, Galton could have weighted each guess according to the ability of that person to estimate such a weight, although quantifying this ability would have been difficult.

---

[1]Reprinted from: Galton, F.: Vox Populi. Nature, **75**, 450–451, ©1907, with permission from Springer Nature.

[2]An interesting discussion of Galton's analysis, including typographical errors, use of the median rather than the mean, and the asymmetric form of the distribution of guesses, can be found in [225].

© The Author(s), under exclusive licence to Springer-Verlag GmbH, DE, part of Springer Nature 2018

D. Fletcher, *Model Averaging*, SpringerBriefs in Statistics, https://doi.org/10.1007/978-3-662-58541-2_1

Over a hundred years later, [215] averaged the results from several cosmological models to estimate the curvature and size of the universe. The models were based on different assumptions about the universe, and were combined using classical Bayesian model averaging, with the weight for each model being the posterior probability that it is true (Sect. 2.2.1). Quite a leap from the weight of an ox to the size of everything.

Model averaging has been used in many other application areas, as illustrated by the references in Table 1.1.

1.2   Benefits of Model Averaging

In much of the theory underlying classical statistical inference, parameter estimation is based on a single model, with this model often being selected as the best from a set of candidate models. The process by which we select this best model is often ignored, leading to point estimates being biased and their precision overestimated [32, 35, 63, 78, 102, 138, 140]. This has been referred to as a quiet scandal [23], and there are likely to be many researchers who are still not aware of this issue.

Model averaging is an approach to estimation that makes some allowance for model uncertainty. In the frequentist framework, it involves calculating a weighted mean of the estimates obtained from each of the candidate models, with the weights reflecting a measure of the potential value of that model for estimation. The model weights might be based on Akaike's information criterion (AIC), cross validation, or the mean squared error (MSE) of the estimate of the parameter of interest (Sect. 3.2). In the Bayesian framework, a model weight is either the posterior probability that the model is true (Sect. 2.2) or is determined using a prediction-based method, such as the Watanabe-Akaike Information Criterion (WAIC) or cross validation. Typically model weights are constrained to lie on the unit simplex, i.e. to be non-negative and sum to one.

From a frequentist perspective, model averaging can also be viewed as a means of achieving a balance between the bias and variance of an estimate, much like model selection. Smaller models will generally provide estimates that have greater bias, whereas larger models will lead to estimates with higher variance. In addition, allowance is made for model uncertainty when calculating a confidence interval, resulting in a wider and more reliable interval than one based on model selection.

Interestingly, some authors in the frequentist domain have focussed solely on achieving a balance between the bias and variance of a model-averaged estimate, while others have considered model averaging solely as a means of allowing for model uncertainty using a model-averaged confidence interval [193].

Table 1.1 References that have used or promoted model averaging, classified by application area

Area of application	References
Econometrics	[4, 37, 38, 49, 50, 59, 61, 62, 71, 81, 96, 115, 122, 130, 133, 134, 144, 145, 160, 169, 179, 202, 224, 229, 230, 241, 242]
Pharmacology	[8, 19, 26, 69, 114, 141, 162, 184, 218, 234, 252, 254]
Meteorology	[41, 95, 123, 124, 154, 178, 198–200, 222, 236]
Hydrology	[57, 65, 163, 175, 185, 212, 223, 239, 249, 255]
Public health	[2, 92, 120, 121, 143, 220, 237, 245, 260, 261]
Ecology	[10, 29, 90, 93, 119, 126, 161, 201, 207, 238]
Environmental risk assessment	[46, 149, 158, 159, 172, 173, 195, 232, 233]
Physics	[11, 66, 72, 167, 168, 215, 235]
Phylogenetics	[20, 82, 129, 176, 177, 186]
Fisheries modelling	[6, 25, 103, 118, 152]
Spatial modelling	[56, 128, 166, 174]
Political science	[14, 74, 156, 157]
Wind-power forecasting	[165, 189, 210]
Meta-analysis	[96, 170, 206]
Clinical trials	[135, 244]
Survival analysis	[209, 221]
Extreme-values	[188, 219]
Forestry science	[171, 214]
Health economics	[104, 105]
Geology	[36, 256]
Climate change	[22, 155]
Bioinformatics	[7, 251]
Environmental modelling	[30, 87]
Remote sensing	[257]
Sensitivity analysis	[250]
Pattern recognition	[231]
Tourism	[226]
Nuclear medicine	[213]
Demographic forecasting	[194]
Psychology	[191]
Soil science	[147]
Entertainment	[127]
Education research	[116]
Chemistry	[88]
Engineering	[15]
Homeland security	[1]

1.3   Examples

We illustrate the use of model averaging with five simple examples. The primary purpose of these is to provide simple numerical illustrations of the methods, rather than detailed insight into their frequentist properties. As our aim is to focus on the key ideas and methods, all of the examples involve small sample sizes and a moderate number of simple models. Many of the ideas and methods will still be of use when we have a large number of more complex models. In order to simplify the discussion in this Chapter, details of the methods used to obtain the model weights, posterior model probabilities, model-averaged estimates and model-averaged confidence intervals are deferred until Chaps. 2 and 3.

1.3.1   Sea Lion Bycatch

The accidental capture and drowning of marine mammals in fishing nets is an important conservation issue in many parts of the world. In order to monitor the situation, some regulatory authorities place observers on fishing vessels, who record the amount of bycatch. The data in Table 1.2 show the number of New Zealand sea lions observed to drown in trawl nets in a fishing area near the Auckland Islands, New Zealand during the 1995–1996 fishing season [148]. The data are classified according to whether the vessel was fishing for scampi, squid or other target species. The total number of tows is also shown, together with the number that were observed.

Suppose we wish to estimate the total number of sea lions killed that season for each of the three types of fishery. One approach is to use a Poisson model with an offset. Thus we assume that

$$Y_i \sim \text{Poisson}\,(\mu_i)\,,$$

$$\log \mu_i = \log n_i + a_i,$$

where Y_i is the number of sea lions killed in the n_i tows observed in fishery i, $\log n_i$ is an offset, and a_i is the effect of target species i ($i = 1, 2, 3$). A natural estimate of θ_i, the total number of sea lions killed by fishery i, is given by

$$\widehat{\theta}_i = N_i \widehat{\gamma}_i,$$

Table 1.2 Bycatch of New Zealand sea lions by trawl nets near the Auckland Islands, New Zealand in the 1995–1996 fishing season

Species	Sea lions	Number of tows	
		Observed	Total
Scampi	3	67	1300
Squid	13	555	4461
Other	1	15	156

Table 1.3 Estimates and 95% confidence intervals for total bycatch (to the nearest integer) of New Zealand sea lions in each of three fisheries

Species		Estimate	Lower	Upper
Scampi	Model 1	35	22	56
	Model 2	58	19	181
	Model-averaged	39	21	118
Squid	Model 1	119	74	192
	Model 2	105	61	180
	Model-averaged	116	69	190
Other	Model 1	4	3	7
	Model 2	10	2	74
	Model-averaged	5	2	35

where N_i is the total number of tows in fishery i, and $\gamma_i = \mu_i/n_i$ is the bycatch rate (sea lions per tow) in fishery i. The following two versions of the model are of interest here:

Model 1 $a_1 = a_2 = a_3$
Model 2 $a_1, a_2, a_3 \in \mathbb{R}$

The estimate of the bycatch rate from model 1 is 0.027 sea lions per tow, whereas model 2 leads to estimates of 0.045, 0.023 and 0.067 for scampi, squid and other species respectively. The AIC model weights (Sect. 3.2.1) are 0.773 and 0.227, for models 1 and 2 respectively. Loosely speaking, these quantify how much more we should value the estimate of bycatch rate from model 1 over those from model 2.

Table 1.3 shows the estimates of total bycatch, together with 95% Wald confidence intervals, plus a model-averaged estimate and 95% confidence interval based on the AIC weights (Sect. 3.4.3). As in any generalised linear model (GLM), it is natural to calculate estimates and confidence intervals on the linear predictor scale, and then transform these back to the original scale. Thus we first calculated a model-averaged estimate and confidence interval for a_i, and converted this to a model-averaged estimate and confidence interval for the total bycatch in fishery i, using the transformation $\theta_i = N_i e^{a_i}$.

The model-averaged estimates and intervals provide a compromise between those obtained from the individual models, with the weighting ensuring that they are closer to those for model 1. Model selection using AIC would lead to inference based on model 1 alone. For the fisheries targeting scampi and other species, the confidence interval obtained from this model is much narrower than that based on model averaging, as use of the best model ignores model uncertainty.

The model-averaged confidence interval reflects the fact that for these two fisheries the estimates and confidence intervals from the two models are quite different. In contrast, for the squid fishery, the model-averaged confidence interval is similar to the two single-model intervals, as the differences between the two models are quite small.

1.3.2   Ecklonia Density

This example is based on a study of the density of a species of seaweed, *Ecklonia radiata*, that was carried out in a fiord on the west coast of Te Wai Pounamu, the south island of New Zealand. Details regarding the study and the dataset can be found in [75]. For simplicity we consider a subset of the data, in which we compare the densities in three zones, classified according to their distance from the mouth of the fiord (0–7 km, 7–10 km, and $\sim$10 km correspond to zones 1 to 3 respectively).

Figure 1.1 shows probability-density histograms summarising the ecklonia densities (individuals per $25\,\mathrm{m}^2$ quadrat) observed in the three zones. This subset of the data involves a total of 102 quadrats (32, 25 and 45 in zones 1 to 3 respectively).

Suppose we wish to estimate the mean density of ecklonia in each of the three zones. An initial analysis suggested that the counts are overdispersed relative to a Poisson model, so we consider use of a negative binomial model, given by

$$\Pr\left(Y_{ij} = y\right) = \frac{\Gamma\left(y + k\right)}{\Gamma\left(y + 1\right)\Gamma\left(k\right)} \left(\frac{k}{\mu_i + k}\right)^k \left(\frac{\mu_i}{\mu_i + k}\right)^y,$$

where Y_{ij} is the density in quadrat j of zone i, μ_i is the mean density in zone i, and k is the dispersion parameter, which is assumed to be the same in each zone ($i = 1, 2, 3$). The following two versions of this model are of interest here:

Model 1 $\mu_1 = \mu_2 = \mu_3$
Model 2 $\mu_1, \mu_2, \mu_3 \in \mathbb{R}$

Model 2 has some lack-of-fit, with a residual deviance of 117.64 on 99 degrees of freedom. This is likely to be due to us not making use of other predictor variables that were available in the original data [75]. In addition, there may be zero-inflation and the value of k might depend on zone. For simplicity of illustration, we do not consider models that allow for these possibilities.

Use of AIC leads to weights of 0.276 and 0.724 for models 1 and 2 respectively, suggesting that we should give more credence to model 2. Table 1.4 shows the estimates of mean density for each model, together with 95% Wald confidence intervals, plus a model-averaged estimate and 95% confidence interval based on the AIC weights. As with the Poisson models for sea lion bycatch (Sect. 1.3.1), it is natural to obtain estimates and confidence intervals on the log-scale, followed by a transformation back to the original scale. The results in Table 1.4 are therefore based on back-transformation of the estimates and confidence intervals for $\log \mu_i$ ($i = 1, 2, 3$).

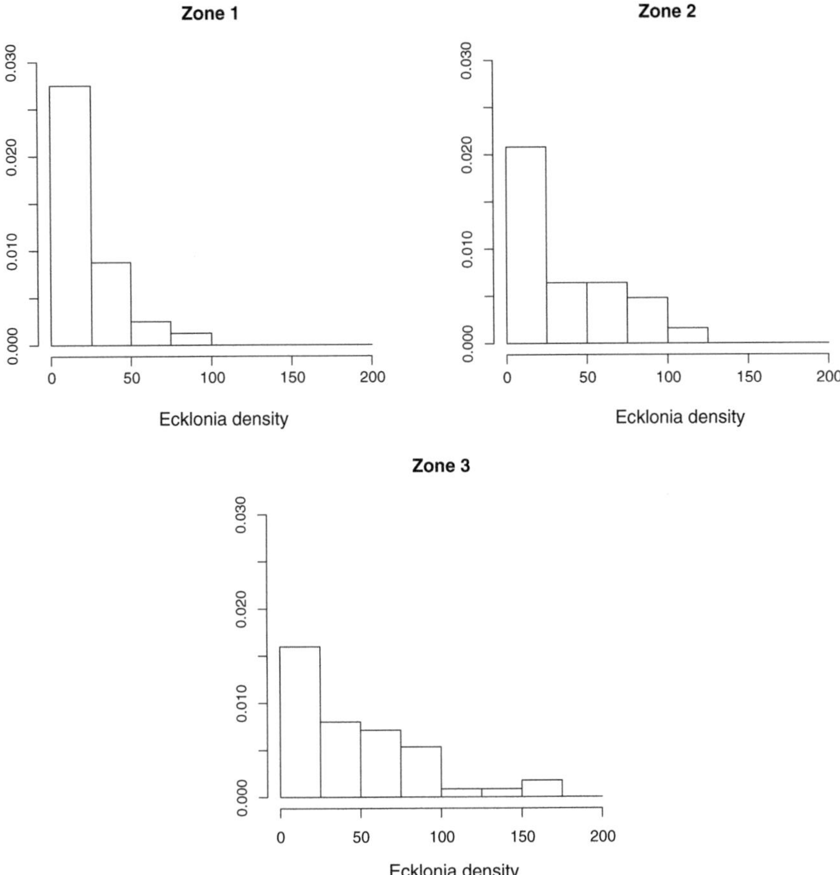

Fig. 1.1 Probability-density histograms of ecklonia densities (individuals per $25\,\text{m}^2$ quadrat) in each of three zones

As in the sea lion example, the model-averaged estimates and intervals provide a compromise between the two models. This example also illustrates an interesting point regarding the choice of weights. For zones 1 and 3 the estimates from the two models are clearly different, while for zone 2 they are almost identical. It would therefore be natural to predict that the bias of the estimate for this zone would be about the same for the two models. This in turn suggests giving more weight to model 1, as a smaller model will provide an estimate with a lower variance. This example illustrates the potential advantage of allowing the choice of model weights to depend upon the parameter of interest (Sect. 3.2.3).

Table 1.4 Estimates and 95% confidence intervals for mean ecklonia density (individuals per quadrat) in each of three zones

		Estimate	Lower	Upper
Zone 1	Model 1	33.5	24.1	46.6
	Model 2	16.9	9.5	30.0
	Model-averaged	20.4	9.9	42.0
Zone 2	Model 1	33.5	24.1	46.6
	Model 2	33.6	17.6	64.0
	Model-averaged	33.6	18.5	61.1
Zone 3	Model 1	33.5	24.1	46.6
	Model 2	45.3	28.0	73.2
	Model-averaged	41.7	25.9	70.7

1.3.3 Water-Uptake in Amphibia

An interesting setting in which to consider use of model averaging is a factorial experiment, in which the set of models is clearly defined. The example we consider was described in [151], and used by [77] as the context for a simulation study of methods for calculating a model-averaged confidence interval. Eight frogs and eight toads were kept in either moist or dry conditions and half were then injected with a water-balance hormone. The response variable was the percentage increase in weight after immersion in water for two hours, with the predictor variables being *species* (frog or toad), *condition* (moist or dry) and *hormone* (yes or no). For simplicity, we will assume that a normal linear model is appropriate. The analysis of variance is shown in Table 1.5.

Table 1.5 Analysis of variance for water-uptake experiment involving the factors *species* (S), *condition* (C) and *hormone* (H)

Source	df	Mean square	F-ratio	p
S	1	514.2	15.0	0.005
C	1	469.8	13.7	0.006
H	1	218.3	6.4	0.036
SC	1	39.4	1.1	0.315
SH	1	165.8	4.8	0.059
CH	1	58.1	1.7	0.229
SCH	1	43.9	1.3	0.291
Error	8	34.3		

Adapted from: Fletcher, D., Dillingham, P.W.: Model-averaged confidence intervals for factorial experiments. Comput. Stat. Data. An. **55**, 3041–3048, ©2011, with permission from Elsevier

Table 1.6 A set of candidate models for the water-uptake experiment, together with AIC weights. Weights larger than 0.1 are shown in bold

Model	AIC weight
Null	0.000
S	0.000
C	0.000
H	0.000
S+C	0.006
S+H	0.001
C+H	0.001
S+C+H	0.030
S+C+SC	0.003
S+H+SH	0.001
C+H+CH	0.000
S+C+H+SC	0.019
S+C+H+SH	**0.161**
S+C+H+CH	0.025
S+C+H+SC+SH	**0.131**
S+C+H+SC+CH	0.018
S+C+H+SH+CH	**0.197**
S+C+H+SC+SH+CH	**0.184**
S+C+H+SC+SH+CH+SCH	**0.222**

Adapted from: Fletcher, D., Dillingham, P.W.: Model-averaged confidence intervals for factorial experiments. Comput. Stat. Data. An. **55**, 3041–3048, ©2011, with permission from Elsevier

Suppose we wish to summarise the results by calculating an estimate and 95% confidence interval for each of the eight treatment-combination means. A common approach is to simply use the full model. As some of the interactions are not statistically significant, this might be inefficient. A natural alternative is to use the estimates and confidence intervals obtained from the best model. For example, we might use the model containing only the main effects, as none of the other terms are statistically significant at the 5% level. Alternatively, we might select the model that has the lowest value of AIC, which turns out to be the full model.

In order to avoid selecting a single best model, we can make use of the AIC weights shown in Table 1.6. The top five models (with weights in bold) include the three main effects plus the species-hormone interaction and have roughly comparable weights, ranging from 0.131 to 0.222.

We can illustrate the benefits of model averaging by considering a subset of the simulation results presented in [77]. Table 1.7 shows the results for the case of two replicates, as in the real data. The coverage rate of a 95% confidence interval for a treatment combination mean has been averaged over all eight combinations, for both the best model and model averaging. The mean width of each interval is also shown, relative to that obtained using the full model. The best model was selected

Table 1.7 Mean coverage rate and relative width of a 95% confidence interval for a treatment combination mean in the water-uptake experiment

Scenario	Mean coverage rate		Mean relative width	
	Best model	Model-averaged	Best model	Model-averaged
Small	0.82	0.95	0.52	0.70
Medium	0.91	0.94	0.81	0.89
Large	0.94	0.94	0.95	0.98

Adapted from: Fletcher, D., Dillingham, P.W.: Model-averaged confidence intervals for factorial experiments. Comput. Stat. Data. An. **55**, 3041–3048, ©2011, with permission from Elsevier

using AIC, and model averaging was performed using the AIC weights in Table 1.6. The results are given for three scenarios, corresponding to the true main effects and interactions all being *small*, *medium* or *large* relative to the error variance [77].

These results show that model averaging can produce an interval that has good coverage and is narrower than one based on the full model. For example, when the true main effects and interactions are small relative to the error variance, the model-averaged interval has perfect coverage and is 30% narrower on average than the interval from the full model. Under this scenario model averaging can be thought of as equivalent to using the full model to analyse an experiment involving $2/0.7^2 = 4.1$ replicates, i.e. the effective replication is just over twice the nominal replication [151]. For the other two scenarios, the model averaged interval provides good coverage and is again narrower on average than the full-model interval. The interval based on the best model is the narrowest on average, but at the expense of having poor coverage, especially when all the main effects and interactions are small.

As we are considering the special case of normal linear models, and the simulations involved the assumption that the full model is the true model,[3] the interval based on that model must have perfect coverage, at the expense of being wider than the model-averaged interval. Thus model averaging has provided a good compromise between coverage and interval-width. This example provides clear evidence against the argument that model averaging is not relevant to the analysis of a designed experiment [28, 217].

1.3.4 Toxicity of a Pesticide

As part of a study to assess the resistance of the tobacco budworm to a pesticide, 20 moths of each sex were assigned to one of six doses of the pesticide. The number of moths that were affected (had uncoordinated movement or were dead) 72 h after treatment is shown in Table 1.8, for each sex and dose [48, 216]. We consider four binomial models, each of the following form:

---

[3] As noted later in this Chapter, we do not regard any model to be a perfect representation of the true date-generating mechanism, but behaving as if a model were true can be useful for inference.

Table 1.8 Results from a toxicity experiment involving the tobacco budworm

Sex	Dose (μg)	Number affected
Male	1	1
	2	4
	4	9
	8	13
	16	18
	32	20
Female	1	0
	2	2
	4	6
	8	10
	16	12
	32	16

Reprinted from: Holloway, J.W.: A comparison of the toxicity of the pyrethroid trans-cypermetherin, with and without the synergist piperonyl butoxide, to adult moths from two strains of Heliothis Virescens. Final-year dissertation, Department of Pure and Applied Zoology, University of Reading, ©1989, with permission from the author

Table 1.9 Models used to analyse the results from a toxicity experiment involving the tobacco budworm

Model	Intercepts	Slopes
1	$a_1 = a_2$	$b_1 = b_2$
2	$a_1, a_2 \in \mathbb{R}$	$b_1 = b_2$
3	$a_1 = a_2$	$b_1, b_2 \in \mathbb{R}$
4	$a_1, a_2 \in \mathbb{R}$	$b_1, b_2 \in \mathbb{R}$

$$Y_{ij} \sim \text{Binomial}\left(20, \pi_{ij}\right),$$

$$\log\left(\frac{\pi_{ij}}{1 - \pi_{ij}}\right) = a_i + b_i x_j,$$

where Y_{ij} is the number of individuals of sex i that were affected at dose d_j, and $x_j = \log_2 d_j$ ($i = 1, 2; j = 1, \ldots, 6$). The choice of a $\log_2$ scale for dose is natural, as it increases in powers of 2. The four models correspond to different assumptions about the intercepts and slopes for males and females, as shown in Table 1.9.

The fit of the largest model (model 4) did not indicate any evidence of overdispersion, so the choice of a binomial distribution for the response variable appears to be reasonable. Figure 1.2 shows the fitted line for each model, together with the observed proportions, separately for each sex.

The AIC weights are 0.004, 0.235, 0.553 and 0.209, for models 1 to 4 respectively. These show evidence in favour of model 3, which allows different slopes for males and

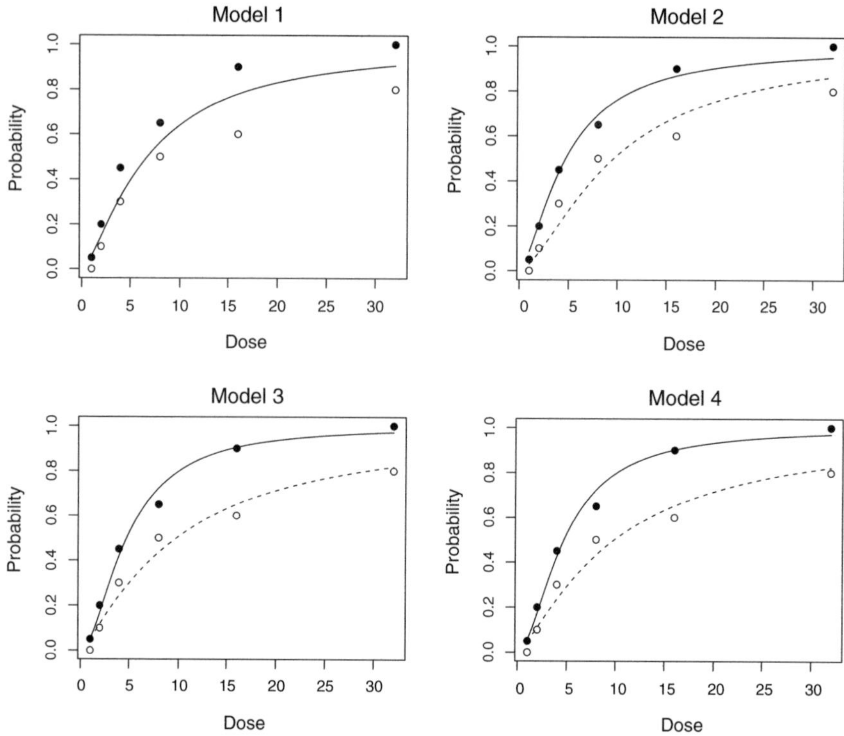

Fig. 1.2 Estimated probability of a tobacco budworm being affected by a specified dose of pesticide, for males (solid) and females (dashed), together with the observed proportions for males (black) and females (white). For model 1, the estimated probability is identical for males and females

females, and moderate support for 2 and 4, both of which allow different intercepts for the two sexes.

As in many toxicity studies, a natural focus of the analysis will be estimation of the dose that leads to a specified probability of an individual being affected. If we denote this probability as π_0, an estimate of the dose required for a particular sex is given by

$$\widehat{d_0} = 2^{\widehat{x_0}},$$

where

$$\widehat{x_0} = \frac{1}{\widehat{b}} \left\{ \log \left(\frac{\pi_0}{1 - \pi_0} \right) - \widehat{a} \right\}$$

and $\widehat{a}, \widehat{b}$ are estimates of the intercept and slope for that sex.

Suppose we are interested in the dose-levels that lead to 50 and 90% of individuals being affected. Table 1.10 shows the estimates and 95% Wald confidence intervals obtained from each model. These are based on back-transformation of the

Table 1.10 Estimates and 95% confidence intervals for the dose-level (μg) of trans-cypermethrin that leads to 50% or 90% of individuals being affected, using each model or model averaging (MA), separately for each sex

Probability affected	Model	Male			Female		
		Estimate	Lower	Upper	Estimate	Lower	Upper
0.5	1	6.7	5.4	8.4	6.7	5.4	8.4
	2	4.7	3.4	6.4	9.6	7.0	13.1
	3	4.8	3.7	6.2	9.8	6.9	14.0
	4	4.7	3.6	6.2	9.9	6.9	14.1
	MA	4.7	3.6	6.3	9.8	6.9	13.8
0.9	1	30.5	19.7	47.2	30.5	19.7	47.2
	2	19.6	12.5	30.9	40.2	24.6	65.6
	3	15.6	10.0	24.2	55.1	27.1	112.4
	4	15.8	9.8	25.5	53.0	24.7	114.0
	MA	16.5	10.2	27.5	50.7	25.6	107.9

corresponding estimate and 95% confidence interval for x_0, and make use of an approximation to the asymptotic standard error of $\widehat{x_0}$ [216].

The model-averaged estimates and 95% confidence intervals are also shown in Table 1.10. These provide a compromise between the estimates and intervals for models 2, 3 and 4, with most weight being given to model 3.

When $\pi_0 = 0.5$, the differences between the estimates for models 2 to 4 are small for both sexes, and the primary effect of the weighting is to discount the results from model 1. This pattern is also evident in the confidence intervals. We would therefore expect the model-averaged estimate and confidence interval to be robust to the choice of weights for models 2 to 4.

When $\pi_0 = 0.9$, some of the estimates and confidence limits lie outside the range of dose-levels used in the study, and in practice we might be wary of such extrapolation. We include this case for illustration only, as the estimate and confidence interval obtained from model 2 are quite different from those for models 3 and 4.

The impact of the difference between the results for the two values of π_0 becomes clearer when we consider different types of model weight in Chaps. 2 and 3.

1.3.5 Assessing the Risk of a Stroke

This example illustrates the advantages of using model averaging to predict the risk of a stroke [221]. Cox proportional hazard models were used to analyse survival data for 4502 individuals, with information on 23 putative risk factors. In order to

Table 1.11 Assigned risk group versus stroke occurrence, for three methods of determining the groups. The total number of individuals are shown for each risk group, together with the percentage that were recorded as having experienced a stroke during the follow-up period

Group	Model averaging		Best PMP model		Stepwise selection	
	% Stroke	Total	% Stroke	Total	% Stroke	Total
Low	0.9	758	1.1	758	1.4	734
Medium	3.0	794	3.3	826	3.4	829
High	7.9	700	7.6	668	7.0	689

Adapted from: Volinsky, C.T., Madigan, D., Raftery, A.E., Kronmal, R.A.: Bayesian model averaging in proportional hazard models: assessing the risk of a stroke. J. Roy. Stat. Soc. C-App. **46**, 433–448 ©1997, with permission from John Wiley & Sons

compare the predictive performance of model selection and model averaging, the data were randomly split into two halves. The first half was used for model selection and model averaging, with the latter using weights based on BIC, the Bayesian information criterion (Sect. 2.2.2). The individuals in the second half of the data were then classified as being at low, medium or high risk of a stroke. Table 1.11 shows the outcomes recorded for these individuals during follow-up, classified by the predictions based on model averaging, and on two choices for the best model: that with the highest posterior model probability (PMP; Sect. 2.2.1) and one obtained using a stepwise backward-elimination process.

Those individuals assigned to the high (low) risk group by model averaging were more (less) likely to have a stroke than those assigned to this group by the other two methods. Thus model averaging was found to be preferable to selection of a best model, in terms of assessing who was at high risk of a stroke.

1.4 When Is Model Averaging Useful?

Model averaging is potentially useful when we are interested in estimation, rather than on the description and understanding of a system [24, 35, 97]. We therefore regard the concept of model-consistency, which is concerned with the asymptotic probability of identifying the true model, as of secondary interest. It is more useful to assess the accuracy of the model-averaged estimate[4] and coverage of the associated confidence interval. The tension that exists between achieving model-consistency and optimal estimation has been considered from a theoretical perspective by [246], who makes the point that when our aim is estimation it is preferable to not even try to find the true model.

We will usually be interested in estimating a function of the model parameters, and it is important that interpretation of this function is the same for all the models,

---

[4]For simplicity, throughout the book we use the term estimate when referring to either an estimator (the method of estimation) or an estimate (the realised value of an estimator), as the meaning should be clear from the context.

the most obvious example being the expected value of the response variable for specified values of the predictor variables. Likewise, it will often not be sensible to model average regression coefficients. For example, suppose we have the following two models for $\mu = E\left(Y \mid x_1, x_2\right)$:

Model 1 $\mu = \beta_0 + \beta_1 x_1$
Model 2 $\mu = \beta_0 + \beta_1 x_1 + \beta_2 x_2$.

Unless the two predictor variables are uncorrelated, the interpretation of β_1 is different for the two models, so it is not appropriate to model average the estimate of this parameter, even though we might think of it as a useful summary of the change in μ that is associated with a unit-change in x_1 [12, 17, 31, 32, 54, 64, 89, 140]. A better alternative is to compare model-averaged estimates of μ for suitably chosen values of both x_1 and x_2 [31].

If a regression coefficient has the same interpretation in each model it is best to average over all the models, rather than exclude those for which the coefficient is zero [5, 28, 140]. This is related to the use of shrinkage in methods such as SCAD penalised regression, ridge regression, the lasso and the elastic net [70, 98, 211, 259]. Shrinkage methods involve one or more tuning parameters, typically estimated using generalised cross validation [240], which determine the amount by which model parameters are shrunk towards zero. They can be thought of as allowing for model uncertainty by simultaneously selecting predictor variables and estimating model parameters. Only recently, however, have confidence intervals based on shrinkage methods been developed that allow for model uncertainty [68]. As pointed out by [228], it would be useful to compare the performance of model averaging and shrinkage methods, for both point and interval estimation.

A useful discussion of the conditions under which we might expect a model-averaged estimate to perform well is given by [60]. If the different models provide estimates that are all negatively biased or all positively biased, the model-averaged estimate will be biased in that direction; it is therefore better for the individual biases to occur in both directions. However, as in model selection, a small amount of bias may be worthwhile if it leads to a large-enough reduction in variance. If the estimates from the different models are strongly positively correlated, the variance of the model-averaged estimate will be greater than if they were uncorrelated. This is related to the concept of model-redundancy [28], which we return to in later chapters.

If two models both have non-negligible weight and give quite different estimates of the parameter(s) of interest, we would expect model averaging to provide a much better measure of uncertainty than that based on either model alone. This has been referred to as "staking out the corners in model-space" [63], and is related to the idea of discrete model averaging [64], in which we use different types of model. A simple example arises when some models differ solely in the distribution of the response variable [28, 244] or the choice of link function [53]. Discrete model averaging often arises in machine learning; in classification, for example, we might use both linear discriminant analysis and random forests to make predictions (Sect. 3.7.6).

It is important to bear in mind that model weights are estimated from the data [34]. In principle, model averaging will be more reliable than model selection, but if the model weights are estimated poorly we may obtain less reliable estimates [60, 86, 164, 181, 182, 197, 253]. Sometimes the primary benefit of model averaging is to simply downweight poorer models, as in the toxicity experiment (Sect. 1.3.4), with the optimal choice of weights for the better models being uncertain [60]. This uncertainty may not matter, however, if the model-averaged estimate is robust to the choice of weights for the better models. An example of this arises when we return to the toxicity experiment in Sect. 3.3.4. There is a need for further work in this area, in terms of assessing both the degree to which the results are robust to the choice of weights, and the impact of estimating the weights imprecisely.

It will often be useful to consider the sensitivity of an estimate to the choice of model, rather than simply average the estimate over a set of models [51, 60, 104]. Likewise, some have argued for the use of continuous model expansion, i.e. use of a single model that contains all other candidate models as special cases, rather than model averaging [64, 84, 247].

We will sometimes find it useful to assume the following framework:

1. We have a finite set of M models
2. Model M is the largest model, within which all other models are nested
3. Model M is true

The actual data-generating mechanism will usually be more complex than any model we can specify, so all of our models will be wrong [28]. However, behaving as if one of the models is true can be a useful basis for inference, in the same way that we traditionally assume the truth of a single model [43, 131, 136, 146]. In addition, of all the models being considered, it is natural to assume that the largest is true, as it is the most complex. This assumption does not imply we should simply base our analysis on the largest model. Even if this model is structurally correct, it might not be the best choice for inference, especially when some of the effect sizes are small (Sect. 1.3.3). Likewise it is still important to check for any lack-of-fit of the largest model; many of the techniques discussed in this book will be of little use if there are strong discrepancies between the data and this model [27, 28, 63, 84, 142].

Although we will often make use of the framework discussed above, there can be advantages in using a method that does not assume the true model is in the model set, notable examples being the Bayesian and frequentist versions of stacking, which use cross validation (Sects. 2.3.2 and 3.2.3).

1.5 Aim of This Book

The aim of this book is to provide a concise overview of model averaging that is accessible to both applied statisticians and researchers with a good background in statistics. Although we consider theoretical developments, our discussion of these will be motivated by a desire to provide researchers with tools and techniques that are straightforward to implement and widely applicable.

As discussed in Sect. 1.3, all of the examples are relatively simple, in order to facilitate discussion of the concepts. As such, they do not illustrate all the issues involved in other settings, such as high-dimensional models. Likewise, I have not attempted to use examples from machine learning and econometrics to illustrate methods used in these areas, as others are much better qualified to provide the relevant context.

The primary focus in this book is the frequentist approach to model averaging, as I believe it is desirable to use a paradigm that is concerned with the repeated-sampling properties of an estimate [52]. There are also practical challenges involved in classical Bayesian model averaging, including the specification of priors, for both parameters and models, and computational issues [13, 43, 44, 47, 67, 71]. Having said that, the Bayesian approach is conceptually elegant, it has the advantage of providing a visual summary of the results in the form of the model-averaged posterior distribution, and the recent development of prediction-based Bayesian model averaging looks promising (Sect. 2.3.2).

The interplay between frequentist and Bayesian approaches to inference has long been a fruitful aspect of statistical theory [16, 21, 83, 187], and this is one of the reasons for including an overview of Bayesian model averaging. We consider this link when discussing different methods in Chaps. 2 and 3. In addition, the Bayesian approach to inference can be thought of as a convenient means of producing methods that may have good frequentist properties [52].

In the frequentist approach to inference it can be a challenge to determine the properties of an estimate across a broad range of scenarios. Asymptotic theory, while often providing more general conclusions and insight than a simulation study, may lack relevance in a realistic setting, where the sample size is finite [107, 227, 228]. Indeed, model averaging is most likely to be useful when the sample size is not large. Even when finite-sample theory is available, it may only apply to a simple setting, such as when we wish to average over a set of normal linear models [110–112]. On the other hand, although simulation studies can provide useful results about sampling properties in realistic settings, they suffer from the difficulty of generalising beyond those settings. There will also be some publication bias associated with such studies, as they are typically carried out in a setting favourable to the method being promoted.

The continual increase in computational power suggests that we should be able to carry out a simulation study to assess the frequentist properties of any model-averaging procedure that we plan to use for a particular analysis. This has the advantage that the simulations can be tailored to the context of the analysis. Having said that, there is scope for broadly-applicable theoretical and simulation-based work on the properties of different model-averaging procedures [60].

The outline of the book is as follows. In Chap. 2 we provide an overview of Bayesian model averaging, both in its own right and with a view to providing some insight into the frequentist approach. In Chap. 3 we consider frequentist methods for calculating a model-averaged estimate and confidence interval. Finally, in Chap. 4 we provide a summary of the key ideas and suggest directions for future research.

1.6   Related Literature

An alternative approach to allowing for model uncertainty involves assessing the sampling properties of an estimate obtained after model selection [9, 18, 39, 40, 94, 100, 102, 108, 109, 180, 196, 248]. This leads to the sampling distribution of an estimate being a mixture distribution, involving the probability that model m is selected and the conditional distribution of the estimate given that model m is selected [67, 137].

A theoretical assessment of the coverage of a confidence interval based on a selected best model has been provided by [106], while [109] proposed an upper bound for the large-sample minimal coverage of such an interval, with a view to indicating when there might be severe under-coverage. In more recent work, [113] have considered the relative contributions of two sources of coverage error for a confidence interval calculated after model selection: selection of the wrong model, and use of the data for both model selection and calculation of the interval. For a simple setting involving two normal linear models, they concluded that selection of the wrong model had the greater effect on coverage error, and that model averaging was therefore preferable to selecting a model using a sample-splitting technique such as cross validation.[5]

In machine learning, use of a shrinkage approach to model averaging has been discussed by [79], who suggested a lasso-type approach to estimating the model weights. Conversely, [192] considers the use of model averaging when performing shrinkage. A combination of shrinkage and model averaging has been considered for logistic regression by [86].

In econometrics there has been a parallel development of methods for combining forecasts from different models. As in model averaging, a weighted mean of the individual forecasts is used. In addition, a criterion such as mean squared error is used to determine an optimal choice of weights, an approach that has also been used in model averaging (Sect. 3.2.3). Unlike model averaging, however, it is sometimes assumed that each forecast is unbiased. An interesting feature of using a forecast-combination is that a model which provides a relatively poor forecast on its own can improve a forecast-combination [85], analogous to the idea of combining weak learners in machine learning [190]. Often the parameters and models are allowed to change over time, and the constraint that the weights be non-negative is sometimes removed [3, 45, 58, 61, 91, 99, 101, 117, 125, 134, 146, 179, 203, 230].

---

[5]Some methods of model averaging, such as stacking (Sects. 2.3.2 and 3.2.3), also make use of sample-splitting.

Ensemble forecasting[6] is related to model averaging, but typically focuses on averaging across different initial conditions or parameter values, rather than different models. It has been used in a range of application areas, including climate change [73, 139, 153, 208], ecology [10], weather prediction [258], and earthquake forecasting [150].

References

1. Abbas, A.E., Tambe, M., von Winterfeldt, D. (eds.): Improving Homeland Security Decisions. Cambridge University Press, Cambridge (2017)
2. Aitkin, M., Liu, C.C., Chadwick, T.: Bayesian model comparison and model averaging for small-area estimation. Ann. Appl. Stat. **3**, 199–221 (2009)
3. Aksu, C., Gunter, S.I.: An empirical analysis of the accuracy of SA, OLS, ERLS and NRLS combination forecasts. Int. J. Forecasting **8**, 27–43 (1992)
4. Amini, S.M., Parmeter, C.F.: Comparisons of model averaging techniques: assessing growth determinants. J. Appl. Econ. **27**, 870–876 (2012)
5. Anderson, D.R.: Model Based Inference in the Life Sciences: A Primer on Evidence. Springer, New York (2008)
6. Anderson, S.C., Cooper, A.B., Jensen, O.P., Minto, C., Thorson, J.T., Walsh, J.C., Afflerbach, J., Dickey-Collas, M., Kleisner, K.M., Longo, C., Osio, G.C., Ovando, D., Mosqueira, I., Rosenberg, A.A., Selig, E.R.: Improving estimates of population status and trend with superensemble models. Fish Fish. **18**, 732–741 (2017)
7. Annest, A., Bumgarner, R.E., Raftery, A.E., Yeung, K.Y.: Iterative Bayesian model averaging: a method for the application of survival analysis to high-dimensional microarray data. BMC Bioinform. **10**, 72 (2009)
8. Aoki, Y., Röshammar, D., Hamrén, B., Hooker, A.C.: Model selection and averaging of nonlinear mixed-effect models for robust phase III dose selection. J. Pharmacokinet. Phar. **44**, 581–597 (2017)
9. Arabatzis, A.A., Gregoire, T.G., Reynolds Jr., M.R.: Conditional interval estimation of the mean following rejection of a two sided test. Commun. Stat.-Theory Methods **18**, 4359–4373 (1989)
10. Arajo, M.B., New, M.: Ensemble forecasting of species distributions. Trends. Ecol. Evol. **22**, 42–47 (2007)
11. Arregui, I.: Bayesian coronal seismology. Adv. Space Res. **61**, 655–672 (2018)
12. Banner, K.M., Higgs, M.D.: Considerations for assessing model averaging of regression coefficients. Ecol. Appl. **27**, 78–93 (2017)
13. Barker, R.J., Link, W.A.: Bayesian multimodel inference by RJMCMC: a Gibbs sampling approach. Am. Stat. **67**, 150–156 (2013)
14. Bartels, L.M.: Specification uncertainty and model averaging. Am. J. Polit. Sci. **41**, 641–674 (1997)
15. Bartz-Beielstein, T., Zaefferer, M., Pham, Q.C.: Optimization via multimodel simulation. Struct. Multidiscipl. Optim. (2018). https://doi.org/10.1007/s00158-018-1934-2
16. Bayarri, M.J., Berger, J.O.: The interplay of Bayesian and frequentist analysis. Stat. Sci. **19**, 58–80 (2004)
17. Berger, J.O., Pericchi, L.R.: Objective Bayesian methods for model selection: introduction and comparison. Lecture Notes-Monograph Series, vol. 38, pp. 135–207. Institute of Mathematical Statistics, Beachwood, Ohio (2001)

---

[6]Not to be confused with the term ensemble method, which is often used in machine learning to describe a technique for model averaging [42].

18. Berk, R., Brown, L., Buja, A., Zhang, K., Zhao, L.: Valid post-selection inference. Ann. Stat. **41**, 802–837 (2013)
19. Bornkamp, B.: Viewpoint: model selection uncertainty, prespecification, and model averaging. Pharm Stat. **14**, 79–81 (2015)
20. Bouckaert, R.R., Drummond, A.J.: bModelTest: Bayesian phylogenetic site model averaging and model comparison. BMC Evol. Biol. **17**, 42 (2017)
21. Box, G.E.P.: Sampling and Bayes' inference in scientific modelling and robustness. J. R. Stat. Soc. Ser. A **143**, 383–430 (1980)
22. Boyd, P.W., Dillingham, P.W., McGraw, C.M., Armstrong, E.A., Cornwall, C.E., Feng, Y.-Y., Hurd, C.L., Gault-Ringold, M., Roleda, M.Y., Timmins-Schiffman, E., Nunn, B.L.: Physiological responses of a Southern Ocean diatom to complex future ocean conditions. Nature Clim. Change **6**, 207–213 (2016)
23. Breiman, L.: The little bootstrap and other methods for dimensionality selection in regression: X-fixed prediction error. J. Am. Stat. Assoc. **87**, 738–754 (1992)
24. Breiman, L.: Statistical modeling: the two cultures (with comments and a rejoinder by the author). Stat. Sci. **16**, 199–231 (2001)
25. Brodziak, J., Piner, K.: Model averaging and probable status of North Pacific striped marlin, Tetrapturus audax. Can. J. Fish. Aquat. Sci. **67**, 793–805 (2010)
26. Buatois, S., Ueckert, S., Frey, N., Retout, S., Mentré, F.: Comparison of model averaging and model selection in dose finding trials analyzed by nonlinear mixed effect models. AAPS J. **20**, 56 (2018)
27. Buckland, S.T., Burnham, K.P., Augustin, N.H.: Model selection: an integral part of inference. Biometrics **53**, 603–618 (1997)
28. Burnham, K.P., Anderson, D.R.: Model Selection and Multimodel Inference: A Practical Information-Theoretic Approach. Springer, New York (2002)
29. Burnham, K.P., Anderson, D.R., Huyvaert, K.P.: AIC model selection and multimodel inference in behavioral ecology: some background, observations, and comparisons. Behav. Ecol. Sociobiol. **65**, 23–35 (2011)
30. Butler, A., Doherty, R.M., Marion, G.: Model averaging to combine simulations of future global vegetation carbon stocks. Environmetrics **20**, 791–811 (2009)
31. Cade, B.S.: Model averaging and muddled multimodel inferences. Ecology **96**, 2370–2382 (2015)
32. Candolo, C., Davison, A.C., Demtrio, C.G.B.: A note on model uncertainty in linear regression. J. R. Stat. Soc. Ser. D (Stat.) **52**, 165–177 (2003)
33. Cane, D., Milelli, M.: Multimodel superensemble technique for quantitative precipitation forecasts in Piemonte region. Nat. Hazard Earth Sys. **10**, 265–273 (2010)
34. Charkhi, A., Claeskens, G., Hansen, B.E.: Minimum mean squared error model averaging in likelihood models. Stat. Sin. **26**, 809–840 (2016)
35. Chatfield, C.: Model uncertainty, data mining and statistical inference. J. Roy. Stat. Soc. Ser. A **158**, 419–466 (1995)
36. Chen, C.-S., Huang, H.-C.: Geostatistical model averaging based on conditional information criteria. Environ. Ecol. Stat. **19**, 23–35 (2012)
37. Chen, J., Li, D., Linton, O., Lu, Z.: Semiparametric ultra-high dimensional model averaging of nonlinear dynamic time series. J. Am. Stat. Assoc. (2017). https://doi.org/10.1080/01621459.2017.1302339
38. Cheng, X., Hansen, B.E.: Forecasting with factor-augmented regression: a frequentist model averaging approach. J. Econ. **186**, 280–293 (2015)
39. Chiou, P., Han, C.-P.: Conditional interval estimation of the exponential location parameter following rejection of a pre-test. Commun. Stat. Theor. Methods **24**, 1481–1492 (1995)
40. Chiou, P.: Interval estimation of scale parameters following a pre-test for two exponential distributions. Comput. Stat. Data. Analy. **23**, 477–489 (1997)
41. Chmielecki, R.M., Raftery, A.E.: Probabilistic visibility forecasting using Bayesian model averaging. Mon. Weather Rev. **139**, 1626–1636 (2011)

42. Clarke, B., Fokoue, E., Zhang, H.H.: Principles and Theory for Data Mining and Machine Learning. Springer, New York (2009)
43. Claeskens, G., Hjort, N.L.: Model Selection and Model Averaging, vol. 330. Cambridge University Press, Cambridge (2008)
44. Claeskens, G.: Focused estimation and model averaging with penalization methods: an overview. Stat. Neerl. **66**, 272–287 (2012)
45. Clemen, R.T.: Combining forecasts: a review and annotated bibliography. Int. J. Forecast. **5**, 559–583 (1989)
46. Clyde, M.: Model uncertainty and health effect studies for particulate matter. Environmetrics **11**, 745–763 (2000)
47. Clyde, M., George, E.I.: Model uncertainty. Stat. Sci. **19**, 81–94 (2004)
48. Collett, D.: Modelling Binary Data. CRC Press, Boca Raton (2002)
49. Cuaresma, J.C., Costantini, M., Hlouskova, J.: Can macroeconomists get rich forecasting exchange rates? No. 176. Vienna University of Economics and Business, Department of Economics, Working Paper (2014)
50. Cotteleer, G., Stobbe, T., van Kooten, G.C.: Bayesian model averaging in the context of spatial hedonic pricing: an application to farmland values. J. Regional. Sci. **51**, 540–557 (2011)
51. Cox, D.R.: Discussion of Draper, D.: Assessment and propagation of model uncertainty (with discussion). J. R. Stat. Soc. Ser. B (Methodol.) **57**, 78 (1995)
52. Cox, D.R.: Principles of Statistical Inference. Cambridge University Press, Cambridge (2006)
53. Czado, C., Raftery, A.E.: Choosing the link function and accounting for link uncertainty in generalized linear models using Bayes factors. Stat. Pap. **47**, 419–442 (2006)
54. Davison, A.C.: Discussion of Chatfield, C.: Model uncertainty, data mining and statistical inference (with discussion). J. R. Stat. Soc. Ser. A **158**, 451–452 (1995)
55. Davison, A.C.: Statistical Models. Cambridge University Press, Cambridge (2003)
56. Dearmon, J., Smith, T.E.: Gaussian process regression and Bayesian model averaging: an alternative approach to modeling spatial phenomena. Geogr. Anal. **48**, 82–111 (2016)
57. Diks, C.G.H., Vrugt, J.A.: Comparison of point forecast accuracy of model averaging methods in hydrologic applications. Stoch. Environ. Res. Risk Assess. **24**, 809–820 (2010)
58. Donaldson, R.G., Kamstra, M.: Forecast combining with neural networks. J. Forecast. **15**, 49–61 (1996)
59. Doppelhofer, G.: Model averaging. In: Durlauf, S.N., Blume, L.E. (eds.) The New Palgrave Dictionary of Economics. Palgrave Macmillan (2008)
60. Dormann, C.F., Calabrese, J.M., GuilleraArroita, G., Matechou, E., Bahn, V., Bartoń, K., Beale, C.M., Ciuti, S., Elith, J., Gerstner, K., Guelat, J., Keil, P., LahozMonfort, J.J., Pollock, L.J., Reineking, B., Roberts, D.R., Schröder, B., Thuiller, W., Warton, D.I., Wintle, B.A., Wood, S.N., Wüest, R.O., Hartig, F.: Model averaging in ecology: a review of Bayesian, information-theoretic, and tactical approaches for predictive inference. Ecol. Monogr. (2018). https://doi.org/10.1002/ecm.1309
61. Drachal, K.: Forecasting spot oil price in a dynamic model averaging framework—have the determinants changed over time? Energ. Econ. **60**, 35–46 (2016)
62. Drachal, K.: Comparison between Bayesian and information-theoretic model averaging: fossil fuels prices example. Energ. Econ. **74**, 208–251 (2018)
63. Draper, D.: Assessment and propagation of model uncertainty. J. R. Stat. Soc. Ser. B (Methodol.) **57**, 45–97 (1995)
64. Draper, D.: Model uncertainty yes, discrete model averaging maybe. Stat. Sci. **14**, 405–409 (1999)
65. Duan, Q., Ajami, N.K., Gao, X., Sorooshian, S.: Multi-model ensemble hydrologic prediction using Bayesian model averaging. Adv. Water Resour. **30**, 1371–1386 (2007)
66. Edeling, W.N., Cinnella, P., Dwight, R.P.: Predictive RANS simulations via Bayesian model-scenario averaging. J. Comput. Phys. **275**, 65–91 (2014)
67. Efron, B.: Estimation and accuracy after model selection. J. Am. Stat. Assoc. **109**, 991–1007 (2014)

68. Ewald, K., Schneider, U.: Uniformly valid confidence sets based on the Lasso. Electron. J. Stat. **12**, 1358–1387 (2018)
69. Faes, C., Aerts, M., Geys, H., Molenberghs, G.: Model averaging using fractional polynomials to estimate a safe level of exposure. Risk. Anal. **27**, 111–123 (2007)
70. Fan, J., Li, R.: Variable selection via nonconcave penalized likelihood and its oracle properties. J. Am. Stat. Assoc. **96**, 1348–1360 (2001)
71. Fernandez, C., Ley, E., Steel, M.F.J.: Benchmark priors for Bayesian model averaging. J. Econ. **100**, 381–427 (2001)
72. Ferrie, C.: Quantum model averaging. New J. Phys. **16**, 093035 (2014)
73. Fildes, R., Kourentzes, N.: Validation and forecasting accuracy in models of climate change. Int. J. Forecast. **27**, 968–995 (2011)
74. Fisher, S.D., Shorrocks, R.: Collective failure? Lessons from combining forecasts for the UK's referendum on EU membership. J. Elections Public Opin. Parties **28**, 59–77 (2018)
75. Fletcher, D., MacKenzie, D., Villouta, E.: Modelling skewed data with many zeros: a simple approach combining ordinary and logistic regression. Environ. Ecol. Stat. **12**, 45–54 (2005)
76. Fletcher, D., Faddy, M.: Confidence intervals for expected abundance of rare species. J. Agric. Biol. Environ. Stat. **12**, 315–324 (2007)
77. Fletcher, D., Dillingham, P.W.: Model-averaged confidence intervals for factorial experiments. Comput. Stat. Data. Anal. **55**, 3041–3048 (2011)
78. Freedman, D.A., Freedman, D.A.: A note on screening regression equations. Am. Stat. **37**, 152–155 (1983)
79. Friedman, J.H., Popescu, B.E.: Predictive learning via rule ensembles. Ann. Appl. Stat. **2**, 916–954 (2008)
80. Galton, F.: Vox populi. Nature **75**, 450–451 (1907)
81. Gao, Y., Long, W., Wang, Z.: Estimating average treatment effect by model averaging. Econ. Lett. **135**, 42–45 (2015)
82. Garamszegi, L.Z., Mundry, R.: Multimodel-inference in comparative analyses. In: Garamszegi, L.Z. (ed.) Modern Phylogenetic Comparative Methods and Their Application in Evolutionary Biology, pp. 305–331. Springer, Heidelberg (2014)
83. Gelman, A., Meng, X.-L., Stern, H.: Posterior predictive assessment of model fitness via realized discrepancies. Stat. Sin. **6**, 733–807 (1996)
84. Gelman, A., Shalizi, C.R.: Philosophy and the practice of Bayesian statistics. Brit. J. Math. Stat. Psychol. **66**, 8–38 (2013)
85. Geweke, J., Amisano, G.: Optimal prediction pools. J. Econ. **164**, 130–141 (2011)
86. Ghosh, D., Yuan, Z.: An improved model averaging scheme for logistic regression. J. Multivariate Anal. **100**, 1670–1681 (2009)
87. Gibbons, J.M., Cox, G.M., Wood, A.T.A., Craigon, J., Ramsden, S.J., Tarsitano, D., Crout, N.M.J.: Applying Bayesian model averaging to mechanistic models: an example and comparison of methods. Environ. Modell. Softw. **23**, 973–985 (2008)
88. Gosink, L.J., Overall, C.C., Reehl, S.M., Whitney, P.D., Mobley, D.L., Baker, N.A.: Bayesian model averaging for ensemble-based estimates of solvation free energies. J. Phys. Chem. B **121**, 3458–3472 (2017)
89. Grömping, U.: Estimators of relative importance in linear regression based on variance decomposition. Am. Stat. **61**, 139–147 (2007)
90. Grueber, C.E., Nakagawa, S., Laws, R.J., Jamieson, I.G.: Multimodel inference in ecology and evolution: challenges and solutions. J. Evol. Bio. **24**, 699–711 (2011)
91. Gunter, S.I.: Nonnegativity restricted least squares combinations. Int. J. Forecast. **8**, 45–59 (1992)
92. Hall, H.I., Song, R., Gerstle III, J.E., Lee, L.M.: Assessing the completeness of reporting of human immunodeficiency virus diagnoses in 2002–2003: capture-recapture methods. Am. J. Epidemiol. **164**, 391–397 (2006)
93. Hamilton, G., McVinish, R., Mengersen, K.: Bayesian model averaging for harmful algal bloom prediction. Ecol. Appl. **19**, 1805–1814 (2009)

94. Han, C.P.: Conditional confidence intervals of regression coefficients following rejection of a preliminary test. In: Ahmed, S.E., Ahsanullah, M., Sinha, B.K. (eds.) Applied Statistical Science, vol. 3, pp. 193–202 (1998)

95. Han, K., Choi, J.T., Kim, C.: Comparison of statistical post-processing methods for probabilistic wind speed forecasting. Asia-Pac. J. Atmos. Sci. **54**, 91–101 (2018)

96. Havranek, T., Horvath, R., Irsova, Z., Rusnak, M.: Cross-country heterogeneity in intertemporal substitution. J. Int. Econ. **96**, 100–118 (2015)

97. Hjort, N.L., Claeskens, G.: Frequentist model average estimators. J. Am. Stat. Assoc. **98**, 879–945 (2003)

98. Hoerl, A.E., Kennard, R.W.: Ridge regression: biased estimation for nonorthogonal problems. Technometrics **12**, 55–67 (1970)

99. Hoogerheide, L., Kleijn, R., Ravazzolo, F., Van Dijk, H.K., Verbeek, M.: Forecast accuracy and economic gains from Bayesian model averaging using time-varying weights. J. Forecast. **29**, 251–269 (2010)

100. Hook, E.B., Regal, R.R.: Validity of methods for model selection, weighting for model uncertainty, and small sample adjustment in capture-recapture estimation. Am. J. Epidemiol. **145**, 1138–1144 (1997)

101. Hsaio, C., Wan, S.K.: Is there an optimal forecast combination? J. Econ. **178**, 294–309 (2014)

102. Hurvich, C.M., Tsai, C.L.: The impact of model selection on inference in linear regression. Am. Stat. **44**, 214–217 (1990)

103. Ianelli, J., Holsman, K.K., Punt, A.E., Aydin, K.: Multi-model inference for incorporating trophic and climate uncertainty into stock assessments. Deep-Sea Res. Pt. **II**(134), 379–389 (2016)

104. Jackson, C.H., Thompson, S.G., Sharples, L.D.: Accounting for uncertainty in health economic decision models by using model averaging. J. R. Stat. Soc. Ser. A **172**, 383–404 (2009)

105. Jackson, C.H., Sharples, L.D., Thompson, S.G.: Structural and parameter uncertainty in Bayesian costeffectiveness models. J. R. Stat. Soc. C-App. **59**, 233–253 (2010)

106. Kabaila, P.: The effect of model selection on confidence regions and prediction regions. Economet. Theor. **11**, 537–537 (1995)

107. Kabaila, P.: On variable selection in linear regression. Economet. Theor. **18**, 913–925 (2002)

108. Kabaila, P.: On the coverage probability of confidence intervals in regression after variable selection. Aust. NZ. J. Stat. **47**, 549–562 (2005)

109. Kabaila, P., Leeb, H.: On the large-sample minimal coverage probability of confidence intervals after model selection. J. Am. Stat. Assoc. **101**, 619–629 (2006)

110. Kabaila, P., Welsh, A.H., Abeysekera, W.: Model-averaged confidence intervals. Scand. J. Stat. **43**, 35–48 (2016)

111. Kabaila, P., Welsh, A.H., Mainzer, R.: The performance of model averaged tail area confidence intervals Commun. Stat-Theor. M. **46**, 10718–10732 (2016)

112. Kabaila, P.: On the minimum coverage probability of model averaged tail area confidence intervals. Can. J. Stat. **46**, 279–297 (2018)

113. Kabaila, P., Mainzer, R.: Two sources of poor coverage of confidence intervals after model selection. Stat. Probabil. Lett. **140**, 185–190 (2018)

114. Kang, S.-H., Kodell, R.L., Chen, J.J.: Incorporating model uncertainties along with data uncertainties in microbial risk assessment. Regul. Toxicol. Pharm. **32**, 68–72 (2000)

115. Kapetanios, G., Labhard, V., Price, S.: Forecasting using Bayesian and information-theoretic model averaging: an application to UK inflation. J. Bus. Econ. Stat. **26**, 33–41 (2008)

116. Kaplan, D., Lee, C.: Optimizing prediction using Bayesian model averaging: examples using large-scale educational assessments. Eval. Rev. (2018). https://doi.org/10.1177/0193841X18761421

117. Kascha, C., Ravazzolo, F.: Combining inflation density forecasts. J. Forecast. **29**, 231–250 (2010)

118. Katsanevakis, S.: Modelling fish growth: model selection, multi-model inference and model selection uncertainty. Fish. Res. **81**, 229–235 (2006)

119. King, R., Brooks, S.P.: Bayesian model discrimination for multiple strata capturerecapture data. Biometrika **89**, 785–806 (2002)
120. King, R., Bird, S.M., Brooks, S.P., Hutchinson, S.J., Hay, G.: Prior information in behavioral capture-recapture methods: demographic influences on drug injectors' propensity to be listed in data sources and their drug-related mortality. Am. J. Epidemiol. **162**, 694–703 (2005)
121. King, R., Bird, S.M., Hay, G., Hutchinson, S.J.: Estimating current injectors in Scotland and their drug-related death rate by sex, region and age-group via Bayesian capture-recapture methods. Stat. Methods Med. Res. **18**, 341–359 (2009)
122. Kitagawa, T., Muris, C.: Model averaging in semiparametric estimation of treatment effects. J. Econ. **193**, 271–289 (2016)
123. Kleiber, W., Raftery, A.E., Baars, J., Gneiting, T., Mass, C.F., Grimit, E.: Locally calibrated probabilistic temperature forecasting using geostatistical model averaging and local Bayesian model averaging. Mon. Weather Rev. **139**, 2630–2649 (2011)
124. Kleiber, W., Raftery, A.E., Gneiting, T.: Geostatistical model averaging for locally calibrated probabilistic quantitative precipitation forecasting. J. Am. Stat. Assoc. **106**, 1291–1303 (2011)
125. Koop, G., Korobilis, D.: Forecasting inflation using dynamic model averaging. Int. Econ. Rev. **53**, 867–886 (2012)
126. Lamon, E.C., Clyde, M.A.: Accounting for model uncertainty in prediction of chlorophyll a in Lake Okeechobee. J. Agr. Biol. Environ. Stat. **5**, 297–322 (2000)
127. Lehrer, S., Xie, T.: Box office buzz: does social media data steal the show from model uncertainty when forecasting for Hollywood? Rev. Econ. Stat. **99**, 749–755 (2017)
128. LeSage, J.P.: Spatial econometric panel data model specification: a Bayesian approach. Spat. Stat.-Neth. **9**, 122–145 (2014)
129. Li, W.L.S., Drummond, A.J.: Model averaging and Bayes factor calculation of relaxed molecular clocks in Bayesian phylogenetics. Mol. Biol. Evol. **29**, 751–761 (2011)
130. Li, D., Linton, O., Lu, Z.: A flexible semiparametric forecasting model for time series. J. Econ. **187**, 345–357 (2015)
131. Link, W., Barker, R.: Model weights and the foundations of multimodel inference. Ecology **87**, 2626–2635 (2006)
132. Little, R.J.: Calibrated bayes. Am. Stat. **60**, 213–223 (2006)
133. Liu, C., Maheu, J.M.: Forecasting realized volatility: a Bayesian model averaging approach. J. Appl. Econ. **24**, 709–733 (2009)
134. Liu, J., Wei, Y., Ma, F., Wahab, M.I.M.: Forecasting the realized range-based volatility using dynamic model averaging approach. Econ. Model. **61**, 12–26 (2017)
135. Longford, N.T.: Synthetic estimators with moderating influence: the carry-over in cross-over trials revisited. Stat. Med. **20**, 3189–3203 (2001)
136. Longford, N.T.: An alternative to model selection in ordinary regression. Stat. Comput. **13**, 67–80 (2003)
137. Longford, N.T.: An alternative analysis of variance. SORT Stat. Oper. Res. T. **32**, 77–92 (2008)
138. Longford, N.T.: "Which model?" is the wrong question. Stat. Neerl. **66**, 237–252 (2012)
139. Lopez, A., Tebaldi, C., New, M., Stainforth, D., Allen, M., Kettleborough, J.: Two approaches to quantifying uncertainty in global temperature changes. J. Clim. **19**, 4785–4796 (2006)
140. Lukacs, P.M., Burnham, K.P., Anderson, D.R.: Model selection bias and Freedmans paradox. Ann. I Stat. Math. **62**, 117–125 (2010)
141. Lunn, D.J.: Automated covariate selection and Bayesian model averaging in population PK/PD models. J. Pharmacokinet. Phar. **35**, 85–100 (2008)
142. Nally, R.M., Duncan, R.P., Thomson, J.R., Yen, J.D.L.: Model selection using information criteria, but is the best model any good? J. Appl. Ecol. **55**, 1441–1444 (2018)
143. Madigan, D., York, J., Allard, D.: Bayesian graphical models for discrete data. Int. Stat. Rev. **63**, 215–232 (1995)
144. Magnus, J.R., Powell, O., Prüfer, P.: A comparison of two model averaging techniques with an application to growth empirics. J. Econ. **154**, 139–153 (2010)

145. Magnus, J.R., Wan, A.T.K., Zhang, X.: Weighted average least squares estimation with non-spherical disturbances and an application to the Hong Kong housing market. Comput. Stat. Data Anal. **55**, 1331–1341 (2011)

146. Magnus, J.R., De Luca, G.: Weighted-average least squares (WALS): a survey. J. Econ. Surv. **30**, 117–148 (2016)

147. Malone, B.P., Minasny, B., Odgers, N.P., McBratney, A.B.: Using model averaging to combine soil property rasters from legacy soil maps and from point data. Geoderma **232**, 34–44 (2014)

148. Manly, B.F.J., Seyb, A., Fletcher, D.J.: Bycatch of sea lions (*Phocarctos Hookeri*) in New Zealand fisheries, 1987/88 to 1995/96, and observer coverage. DOC Science Internal Series Number 42. Department of Conservation, Wellington (2002)

149. Martin, M.A., Roberts, S.: Bootstrap model averaging in time series studies of particulate matter air pollution and mortality. J. Expo. Sci. Environ. Epid. **16**, 242–250 (2005)

150. Marzocchi, W.J., Zechar, D., Jordan, T.H.: Bayesian forecast evaluation and ensemble earthquake forecasting. B. Seismol. Soc. Am. **102**, 2574–2584 (2012)

151. Mead, R.: The Design of Experiments: Statistical Principles for Practical Applications. Cambridge University Press, Cambridge (1988)

152. Millar, C.P., Jardim, E., Scott, F., Osio, G.C., Mosqueira, I., Alzorriz, N.: Model averaging to streamline the stock assessment process. ICES J. Mar. Sci. **72**, 93–98 (2015)

153. Min, S.-K., Simonis, D., Hense, A.: Probabilistic climate change predictions applying Bayesian model averaging. Philos. Trans. R. Soc. Lond. A Math. Phys. Eng. Sci. **365**, 2103–2116 (2007)

154. Möller, A., Lenkoski, A., Thorarinsdottir, T.L.: Multivariate probabilistic forecasting using ensemble Bayesian model averaging and copulas. Q. J. R. Meteor. Soc. **139**, 982–991 (2013)

155. Moftakhari, H., AghaKouchak, A., Sanders, B.F., Matthew, R.A., Mazdiyasni, O.: Translating uncertain sea-level projections into infrastructure impacts using a Bayesian framework. Geophys. Res. Lett. **44**, 11914–11921 (2017)

156. Montgomery, J.M., Nyhan, B.: Bayesian model averaging: theoretical developments and practical applications. Polit. Anal. **18**, 245–270 (2010)

157. Montgomery, J.M., Hollenbach, F.M., Ward, M.D.: Improving predictions using ensemble Bayesian model averaging. Polit. Anal. **20**, 271–291 (2012)

158. Moon, H., Kim, H.-J., Chen, J.J., Kodell, R.L.: Model averaging using the Kullback information criterion in estimating effective doses for microbial infection and illness. Risk Anal. **25**, 1147–1159 (2005)

159. Moon, H., Kim, S.B., Chen, J.J., George, N.I., Kodell, R.L.: Model uncertainty and model averaging in the estimation of infectious doses for microbial pathogens. Risk Anal. **33**, 220–231 (2013)

160. Moulton, B.R.: A Bayesian approach to regression selection and estimation, with application to a price index for radio services. J. Econ. **49**, 169–193 (1991)

161. Nakagawa, S., Freckleton, R.P.: Model averaging, missing data and multiple imputation: a case study for behavioural ecology. Behav. Ecol. Sociobiol. **65**, 103–116 (2011)

162. Namata, H., Aerts, M., Faes, C., Teunis, P.: Model averaging in microbial risk assessment using fractional polynomials. Risk Anal. **28**, 891–905 (2008)

163. Neuman, S.P.: Maximum likelihood Bayesian averaging of uncertain model predictions. Stoch. Environ. Res. Risk Assess. **17**, 291–305 (2003)

164. Nguefack-Tsague, G.: On optimal weighting scheme in model averaging. Am. J. Appl. Math. Stat. **2**, 150–156 (2014)

165. Nielsen, H.A., Nielsen, T.S., Madsen, H., Pindado, M.J.S.I., Marti, I.: Optimal combination of wind power forecasts. Wind Energy **10**, 471–482 (2007)

166. Oberdabernig, D.A., Humer, S., Crespo Cuaresma, J.: Democracy, geography and model uncertainty. Scot. J. Polit. Econ. **2017**(65), 154–185 (2018)

167. Parkinson, D., Liddle, R.A.: Application of Bayesian model averaging to measurements of the primordial power spectrum. Phys. Rev. D. **82**, 103533 (2010)

168. Parkinson, D., Liddle, A.R.: Bayesian model averaging in astrophysics: a review. Stat. Anal. Data. Min. **6**, 3–14 (2013)

169. Pesaran, M.H., Schleicher, C., Zaffaroni, P.: Model averaging in risk management with an application to futures markets. J. Empir. Financ. **16**, 280–305 (2009)
170. Philips, A.Q.: Seeing the forest through the trees: a meta-analysis of political budget cycles. Public Choice **168**, 313–341 (2016)
171. Picard, N., Henry, M., Mortier, F., Trotta, C., Saint-André, L.: Using Bayesian model averaging to predict tree aboveground biomass in tropical moist forests. Forest Sci. **58**, 15–23 (2012)
172. Piegorsch, W.W., An, L., Wickens, A.A., Webster West, R., Pea, E.A., Wu, W.: Information-theoretic modelaveraged benchmark dose analysis in environmental risk assessment. Environmetrics **24**, 143–157 (2013)
173. Piegorsch, W.W.: Model uncertainty in environmental dose-response risk analysis. Stat. Publ. Pol. **1**, 78–85 (2014)
174. Piribauer, P.: Heterogeneity in spatial growth clusters. Empir. Econ. **51**, 659–680 (2016)
175. Poeter, E., Anderson, D.: Multimodel ranking and inference in ground water modeling. Groundwater **43**, 597–605 (2005)
176. Posada, D., Buckley, T.R.: Model selection and model averaging in phylogenetics: advantages of Akaike information criterion and Bayesian approaches over likelihood ratio tests. Syst. Biol. **53**, 793–808 (2004)
177. Posada, D.: jModelTest: phylogenetic model averaging. Mol. Biol. Evol. **25**, 1253–1256 (2008)
178. Raftery, A.E., Gneiting, T., Balabdaoui, F., Polakowski, M.: Using Bayesian model averaging to calibrate forecast ensembles. Mon. Weather Rev. **133**, 1155–1174 (2005)
179. Raftery, A.E., Krn, M., Ettler, P.: Online prediction under model uncertainty via dynamic model averaging: Application to a cold rolling mill. Technometrics **52**, 52–66 (2010)
180. Regal, R.R., Hook, E.B.: The effects of model selection on confidence intervals for the size of a closed population. Stat. Med. **10**, 717–721 (1991)
181. Richards, S.A.: Testing ecological theory using the information-theoretic approach: examples and cautionary results. Ecology **86**, 2805–2814 (2005)
182. Richards, S.A., Whittingham, M.J., Stephens, P.A.: Model selection and model averaging in behavioural ecology: the utility of the IT-AIC framework. Behav. Ecol. Sociobiol. **65**, 77–89 (2011)
183. Ripley, B.D.: Selecting amongst large classes of models. In: Adams, N., Crowder, M., Hand, D.J., Stephens, D. (eds.) Methods and Models in Statistics: in Honor of Professor John Nelder, FRS, pp. 155–170. Imperial College Press, London (2004)
184. Ritz, C., Gerhard, D., Hothorn, L.A.: A unified framework for benchmark dose estimation applied to mixed models and model averaging. Stat. Biopharm. Res. **5**, 79–90 (2013)
185. Rojas, R., Feyen, L., Dassargues, A.: Conceptual model uncertainty in groundwater modeling: combining generalized likelihood uncertainty estimation and Bayesian model averaging. Water Resour. Res. **44**, W12418 (2008)
186. Ronquist, F., Teslenko, M., van der Mark, P., Ayres, D.L., Darling, A., Höhna, S., Larget, B., Liu, L., Suchard, M.A., Huelsenbeck, J.P.: MrBayes 3.2: efficient Bayesian phylogenetic inference and model choice across a large model space. Syst. Biol. **61**, 539–542 (2012)
187. Rubin, D.B.: Bayesianly justifiable and relevant frequency calculations for the applied statistician. Ann. Stat. **12**, 1151–1172 (1984)
188. Sabourin, A., Naveau, P., Fougres, A.-L.: Bayesian model averaging for multivariate extremes. Extremes **16**, 325–350 (2013)
189. Snchez, I.: Adaptive combination of forecasts with application to wind energy. Int. J. Forecast. **24**, 679–693 (2008)
190. Schapire, R.E.: The strength of weak learnability. Mach. Learn. **5**, 197–227 (1990)
191. Scheibehenne, B., Gronau, Q.F., Jamil, T., Wagenmakers, E.-J.: Fixed or random? A resolution through model averaging: reply to Carlsson, Schimmack, Williams, and Bürkner (2017). Psychol. Sci. **28**, 1698–1701 (2017)
192. Schomaker, M.: Shrinkage averaging estimation. Stat. Pap. **53**, 1015–1034 (2012)
193. Schomaker, M., Heumann, C.: When and when not to use optimal model averaging (2018). arXiv preprint: arXiv:1802.04589

194. Shang, H.L., Wisniowski, A., Bijak, J., Smith, P.W., Raymer, J.: Bayesian functional models for population forecasting. Working Paper 12.1, Statistical Office of the European Union (2013)
195. Shao, K., Gift, J.S.: Model uncertainty and Bayesian model averaged benchmark dose estimation for continuous data. Risk Anal. **34**, 101–120 (2014)
196. Shen, X., Huang, H.-C., Ye, J.: Inference after model selection. J. Am. Stat. Assoc. **99**, 751–762 (2004)
197. Shen, X., Huang, H.-C.: Optimal model assessment, selection, and combination. J. Am. Stat. Assoc. **101**, 554–568 (2006)
198. Sloughter, J.M.L., Raftery, A.E., Gneiting, T., Fraley, C.: Probabilistic quantitative precipitation forecasting using Bayesian model averaging. Mon. Weather Rev. **135**, 3209–3220 (2007)
199. Sloughter, J.M.L., Gneiting, T., Raftery, A.E.: Probabilistic wind speed forecasting using ensembles and Bayesian model averaging. J. Am. Stat. Assoc. **105**, 25–35 (2010)
200. Sloughter, J.M.L., Gneiting, T., Raftery, A.E.: Probabilistic wind vector forecasting using ensembles and Bayesian model averaging. Mon. Weather Rev. **141**, 2107–2119 (2013)
201. Stanley, T.R., Burnham, K.P.: Information-theoretic model selection and model averaging for closed-population capture-recapture studies. Biometrical J. **40**, 475–494 (1918)
202. Steel, M.F.J.: Bayesian model averaging and forecasting. Bull. EU US Inflat. Macroecon. Anal. **200**, 30–41 (2011)
203. Stock, J.H., Watson, M.W.: Forecasting with many predictors. In: Elliott, C.G.G., Timmermann, A. (eds.) Handbook of Economic Forecasting. Elsevier (2006)
204. Stone, M.: Comments on model selection criteria of Akaike and Schwarz. J. R. Stat. Soc. B. Met. **41**, 276–278 (1979)
205. Surowiecki, J.: The Wisdom of Crowds. Anchor Books (2005)
206. Sutton, A.J., Abrams, K.R.: Bayesian methods in meta-analysis and evidence synthesis. Stat. Methods Med. Res. **10**, 277–303 (2001)
207. Symonds, M.R.E., Moussalli, A.: A brief guide to model selection, multimodel inference and model averaging in behavioural ecology using Akaikes information criterion. Behav. Ecol. Sociobiol. **65**, 13–21 (2011)
208. Tebaldi, C., Smith, R.L., Nychka, D., Mearns, L.O.: Quantifying uncertainty in projections of regional climate change: a Bayesian approach to the analysis of multimodel ensembles. J. Climate **18**, 1524–1540 (2005)
209. Thamrin, S.A., McGree, J.M., Mengersen, K.L.: Modelling survival data to account for model uncertainty: a single model or model averaging? SpringerPlus **2**, 665 (2013)
210. Thordarson, F.Ö., Madsen, H., Nielsen, H.A., Pinson, P.: Conditional weighted combination of wind power forecasts. Wind Energy **13**, 751–763 (2010)
211. Tibshirani, R.: Regression shrinkage and selection via the lasso: a retrospective. J. R. Stat. Soc. B Met. **73**, 273–282 (2011)
212. Tsai, F.T.C., Li, X.: Inverse groundwater modeling for hydraulic conductivity estimation using Bayesian model averaging and variance window. Water Resour. Res. **44**, W09434 (2008)
213. Turkheimer, F.E., Hinz, R., Cunningham, V.J.: On the undecidability among kinetic models: from model selection to model averaging. J. Cerebr. Blood F. Met. **23**, 490–498 (2003)
214. van Oijen, M., Reyer, C., Bohn, F.J., Cameron, D.R., Deckmyn, G., Flechsig, M., Härkönen, S., Hartig, F., Huth, A., Kiviste, A., Lasch, P., Mäkelä, A., Mette, T., Minunno, F., Rammer, W.: Bayesian calibration, comparison and averaging of six forest models, using data from Scots pine stands across Europe. Forest Ecol. Manag. **289**, 255–268 (2013)
215. Vardanyan, M., Trotta, R., Silk, J.: Applications of Bayesian model averaging to the curvature and size of the Universe. Mon. Not. R. Astron. Soc. **413**, L91–L95 (2011)
216. Venables, W.N., Ripley, B.D.: Modern Applied Statistics with S. Springer, New York (2002)
217. Ver Hoef, J.M., Boveng, P.L.: Iterating on a single model is a viable alternative to multimodel inference. J. Wildl. Manag. **79**, 719–729 (2015)
218. Verrier, D., Sivapregassam, S., Solente, A.-C.: Dose-finding studies, MCP-Mod, model selection, and model averaging: two applications in the real world. Clin. Trials **11**, 476–484 (2014)

219. Vettori, S., Huser, R., Segers, J., Genton, M.G.: Bayesian model averaging over tree-based dependence structures for multivariate extremes (2017). arXiv preprint: arXiv:1705.10488
220. Viallefont, V., Raftery, A.E., Richardson, S.: Variable selection and Bayesian model averaging in casecontrol studies. Stat. Med. **20**, 3215–3230 (2001)
221. Volinsky, C.T., Madigan, D., Raftery, A.E., Kronmal, R.A.: Bayesian model averaging in proportional hazard models: assessing the risk of a stroke. J. R. Stat. Soc. C-App. **46**, 433–448 (1997)
222. Vrugt, J.A., Clark, M.P., Diks, C.G.H., Robinson, B.A.: Multi-objective calibration of forecast ensembles using Bayesian model averaging. Geophys. Res. Lett. **33**, L19817 (2006)
223. Vrugt, J.A., Robinson, B.A.: Treatment of uncertainty using ensemble methods: comparison of sequential data assimilation and Bayesian model averaging. Water Resour. Res. **43**, W01411 (2007)
224. Wagner, M., Hlouskova, J.: Growth regressions, principal components augmented regressions and frequentist model averaging. Jahrb. Natl. Stat. **235**, 642–662 (2015)
225. Wallis, K.F.: Revisiting Francis Galton's forecasting competition. Stat. Sci. **29**, 420–424 (2014)
226. Wan, A.T.K., Zhang, X.: On the use of model averaging in tourism research. Ann. Tourism Res. **36**, 525–532 (2009)
227. Wan, A.T.K., Zhang, X., Zou, G.: Least squares model averaging by Mallows criterion. J. Econ. **156**, 277–283 (2010)
228. Wang, H., Zhang, X., Zou, G.: Frequentist model averaging estimation: a review. J. Syst. Sci. Complex. **22**, 732–748 (2009)
229. Wang, C., Nishiyama, Y.: Volatility forecast of stock indices by model averaging using high-frequency data. Int. Rev. Econ. Financ. **40**, 324–337 (2015)
230. Wang, Y., Ma, F., Wei, Y., Wu, C.: Forecasting realized volatility in a changing world: a dynamic model averaging approach. J. Bank. Financ. **64**, 136–149 (2016)
231. Webb, A.R.: Statistical Pattern Recognition. Wiley, Chichester (2003)
232. Wheeler, M.W., Bailer, A.J.: Properties of modelaveraged BMDLs: a study of model averaging in dichotomous response risk estimation. Risk Anal. **27**, 659–670 (2007)
233. Wheeler, M.W., Bailer, A.J.: Comparing model averaging with other model selection strategies for benchmark dose estimation. Environ. Ecol. Stat. **16**, 37–51 (2009)
234. Whitney, M., Ryan, L.: Quantifying dose-response uncertainty using Bayesian model averaging. In: Cooke, R.M. (ed.) Uncertainty modeling in dose response: bench testing environmental toxicity, vol. 74, pp. 165–179 (2009)
235. Williams, M.: A novel approach to the bias-variance problem in bump hunting. J. Instrum. **12**, P09034 (2017)
236. Wilson, L.J., Beauregard, S., Raftery, A.E., Verret, R.: Calibrated surface temperature forecasts from the Canadian ensemble prediction system using Bayesian model averaging. Mon. Weather. Rev. **135**, 1364–1385 (2007)
237. Wilson, A., Zigler, C.M., Patel, C.J., Dominici, F.: Model-averaged confounder adjustment for estimating multivariate exposure effects with linear regression. Biometrics (2018). https://doi.org/10.1111/biom.12860
238. Wintle, B.A., McCarthy, M.A., Volinsky, C.T., Kavanagh, R.P.: The use of Bayesian model averaging to better represent uncertainty in ecological models. Conserv. Biol. **17**, 1579–1590 (2003)
239. Wöhling, T., Vrugt, J.A.: Combining multiobjective optimization and Bayesian model averaging to calibrate forecast ensembles of soil hydraulic models. Water Resour. Res. **44**, W12432 (2008)
240. Wood, S.N.: Generalized additive models: an introduction with R. Chapman and Hall/CRC (2017)
241. Wright, J.H.: Bayesian model averaging and exchange rate forecasts. J. Econ. **146**, 329–341 (2008)
242. Wright, J.H.: Forecasting US inflation by Bayesian model averaging. J. Forecasting **28**, 131–144 (2009)

243. Xie, M.g., Singh, K.: Confidence distribution, the frequentist distribution estimator of a parameter: a review. Int. Stat. Rev. **81**, 3–39 (2013)

244. Xu, R., Mehrotra, D.V., Shaw, P.A.: Incorporating baseline measurements into the analysis of crossover trials with time to event endpoints. Stat. Med. (2018). https://doi.org/10.1002/sim.7834

245. Yamana, T.K., Kandula, S., Shaman, J.: Individual versus superensemble forecasts of seasonal influenza outbreaks in the United States. PLoS Comput. Biol. **13**, e1005801 (2017)

246. Yang, Y.: Can the strengths of AIC and BIC be shared? A conflict between model indentification and regression estimation. Biometrika **92**, 937–950 (2005)

247. Yao, Y., Vehtari, A., Simpson, D., Gelman, A.: Using stacking to average Bayesian predictive distributions. Bayesian Analysis (2018). https://doi.org/10.1214/17-BA1091

248. Ye, J.: On measuring and correcting the effects of data mining and model selection. J. Am. Stat. Assoc. **93**, 120–131 (1998)

249. Ye, M., Pohlmann, K.F., Chapman, J.B., Pohll, G.M., Reeves, D.M.: A model-averaging method for assessing groundwater conceptual model uncertainty. Groundwater **48**, 716–728 (2010)

250. Ye, M., Hill, M.C.: Global sensitivity analysis for uncertain parameters, models, and scenarios. In: Petropoulos, G.P., Srivastava, P.K. (eds.) Sensitivity analysis in earth observation modelling. Elsevier (2017)

251. Yeung, K.Y., Bumgarner, R.E., Raftery, A.E.: Bayesian model averaging: development of an improved multi-class, gene selection and classification tool for microarray data. Bioinformatics **21**, 2394–2402 (2005)

252. Yin, G., Yuan, Y.: Bayesian model averaging continual reassessment method in phase I clinical trials. J. Am. Stat. Assoc. **104**, 954–968 (2009)

253. Yuan, Z., Yang, Y.: Combining linear regression models. J. Am. Stat. Assoc. **100**, 1202–1214 (2005)

254. Yuan, Y., Yin, G.: Robust EM continual reassessment method inoncology dose finding. J. Am. Stat. Assoc. **106**, 818–831 (2011)

255. Zhang, X., Srinivasan, R., Bosch, D.: Calibration and uncertainty analysis of the SWAT model using genetic algorithms and Bayesian model averaging. J. Hydrol. **374**, 307–317 (2009)

256. Zhang, J., Huang, H.W., Juang, C.H., Su, W.W.: Geotechnical reliability analysis with limited data: consideration of model selection uncertainty. Eng. Geo. **181**, 27–37 (2014)

257. Zhao, K., Valle, D., Popescu, S., Zhang, X., Mallick, B.: Hyperspectral remote sensing of plant biochemistry using Bayesian model averaging with variable and band selection. Remote Sens. Environ. **132**, 102–119 (2013)

258. Zhou, B., Du, J.: Fog prediction from a multimodel mesoscale ensemble prediction system. Wea. Forecasting **25**, 303–322 (2010)

259. Zou, H., Hastie, T.: Regularization and variable selection via the elastic net. J. R. Stat. Soc. B Met. **67**, 301–320 (2005)

260. Zwane, E., van der Heijden, P.: Population estimation using the multiple system estimator in the presence of continuous covariates. Stat. Model. **5**, 39–52 (2005)

261. Zwane, E., van der Heijden, P.G.: Capture-recapture studies with incomplete mixed categorical and continuous covariates. J. Data Sci **6**, 557–572 (2008)

Chapter 2
Bayesian Model Averaging

Abstract We provide an overview of Bayesian model averaging (BMA), starting with a summary of the mathematics associated with classical BMA, including the calculation of posterior model probabilities and the choice of priors for both the models and the model parameters. We also consider prediction-based approaches to BMA and argue that these are preferable to the classical approach. Use of BMA is illustrated by two examples involving real data. We finish with a discussion of the advantages and disadvantages of BMA.

2.1 Introduction

In principle, Bayesian model averaging (BMA) is quite natural, as the Bayesian framework for data analysis can readily incorporate both parameter uncertainty and model uncertainty [73, 87]. Instead of the posterior distribution for a parameter being based on a single model, we calculate a weighted combination of the posterior distributions from the different models.

In the classical version of BMA the weight for a model is the posterior probability that it is true, as we assume that one of the models is true. Recently, an explicitly prediction-based version of BMA has been proposed by [181]. This uses cross validation to determine a set of optimal weights, and is a Bayesian version of the frequentist method known as stacking (Sect. 3.2.3).[1]

Classical BMA is concerned with model-consistency (identification of the true model), whereas prediction-based BMA has the advantage of being directly concerned with estimation. We initially focus on classical BMA, as it has received the vast majority of attention in the BMA literature.

---

[1] The assumption that one of the models is true is sometimes referred to in the Bayesian literature as the $\mathcal{M}$-closed framework [13], and it is natural to use classical BMA within this framework. The $\mathcal{M}$-open framework assumes that the true data-generating mechanism is not in the model set, and prediction-based BMA is preferable in this framework.

© The Author(s), under exclusive licence to Springer-Verlag GmbH, DE, part of Springer Nature 2018
D. Fletcher, *Model Averaging*, SpringerBriefs in Statistics,
https://doi.org/10.1007/978-3-662-58541-2_2

2.2 Classical BMA

Once the priors are specified for both the models and the parameters in each model, classical BMA involves a well-defined procedure for obtaining the model-averaged posterior for any parameter of interest. As with other Bayesian methods, the apparent simplicity of classical BMA can bely difficulties in its implementation [8, 32]. The choice of priors can be problematic, as can the computations required to obtain the model-averaged posterior.

Suppose $y = (y_1, \ldots, y_n)^\top$ contains the values of the response variable, and we wish to average over a set of M nested models, with β_m being the vector of p_m parameters in model m. As the notation implies, β_m will often be a set of regression coefficients, but it may include other types of parameter, such as the error variance in a normal linear model.[2] Suppose we are interested in estimating a scalar parameter θ. The model-averaged posterior for θ is given by

$$p\left(\theta\middle|y\right) = \sum_{m=1}^{M} p\left(m\middle|y\right) p\left(\theta\middle|y, m\right), \tag{2.1}$$

where $p\left(m\middle|y\right)$ is the posterior probability that model m is true, and $p\left(\theta\middle|y, m\right)$ is the posterior for θ when we assume model m is true.[3] Thus $p\left(\theta\middle|y\right)$ is a weighted combination of the posterior distributions obtained from the different models, the weights being the posterior model probabilities. Using Bayes' theorem, the posterior probability for model m is given by

$$p\left(m\middle|y\right) \propto p\left(m\right) p\left(y\middle|m\right), \tag{2.2}$$

where $p\left(m\right)$ and $p\left(y\middle|m\right)$ are the prior probability, and the marginal (integrated) likelihood, for model m, with

$$p\left(y\middle|m\right) = \int p\left(y\middle|\beta_m, m\right) p\left(\beta_m\middle|m\right) d\beta_m, \tag{2.3}$$

where $p\left(\beta_m\middle|m\right)$ is the prior for β_m and $p\left(y\middle|\beta_m, m\right)$ is the likelihood under model m. When the support of β_m is discrete, the integral in (2.3) is replaced by a summation.

The posterior probability for model m can also be written as

$$p\left(m\middle|y\right) \propto p\left(m\right) B_m,$$

---

[2]Whether or not we include any scale and shape parameters in our definition of p_m will typically make no difference to the model weights, as long as every model includes these parameters.

[3]For simplicity, we avoid attaching a subscript to $p(\cdot)$ in order to fully specify the distribution. The exception to this will be in Sect. 2.5, where we need to make a clear distinction.

where

$$B_m = \frac{p\left(y|m\right)}{p\left(y|1\right)}$$

is the Bayes factor for comparing model m and model 1, the latter being an arbitrary reference model [114, 146]. It is well known that Bayes factors can be sensitive to the prior distributions for the parameters, even when n is large [8, 97]. An extreme case arises when one or more of the priors is improper, as this can lead to the marginal likelihood in (2.3) not being well defined [73, 161].

Two natural summaries of the model-averaged posterior for θ are the mean and variance, given by

$$\mathrm{E}\left(\theta|y\right) = \sum_{m=1}^{M} p\left(m|y\right) \mathrm{E}\left(\theta|y, m\right) \tag{2.4}$$

and

$$\mathrm{var}\left(\theta|y\right) = \sum_{m=1}^{M} p\left(m|y\right) \left[\mathrm{var}\left(\theta|y, m\right) + \left\{\mathrm{E}\left(\theta|y, m\right) - \mathrm{E}\left(\theta|y\right)\right\}^2\right], \tag{2.5}$$

where $\mathrm{E}\left(\theta|y, m\right)$ and $\mathrm{var}\left(\theta|y, m\right)$ are the posterior mean and variance for θ under model m. Thus the model-averaged posterior variance is influenced by both the parameter uncertainty associated with each model and the between-model variability in the posterior mean.

We can also use the model-averaged posterior to calculate a central $100\left(1 - 2\alpha\right)\%$ credible interval for θ, given by $[\theta_L, \theta_U]$, where

$$\int_{-\infty}^{\theta_L} p\left(\theta|y\right) d\theta = \int_{\theta_U}^{\infty} p\left(\theta|y\right) d\theta = \alpha. \tag{2.6}$$

An alternative choice would be to use the highest posterior density $100\left(1 - 2\alpha\right)\%$ credible region, i.e. the region of values for θ that contain $100\left(1 - 2\alpha\right)\%$ of the posterior probability and for which the posterior density is never lower than outside the region. However, the central credible interval is easier to compute than the highest posterior density region, and has the advantage that the limits can be interpreted as quantiles of the posterior. In the examples we therefore use central credible intervals.

A by-product of classical BMA is the ability to calculate a posterior inclusion-probability (PIP) for each predictor variable, i.e. the sum of the posterior probabilities for all the models that include that variable [7, 12, 36]. Some authors have suggested that these provide a useful summary of the relative importance of each predictor variable. However, as they are influenced by the choice of model set, the importance of a predictor variable can be exaggerated by including many models containing that variable [66]. In addition, a more useful summary of relative importance can be obtained by comparing model-averaged posterior distributions for the expected value of the response variable for a suitable set of values of the predictor variables (Sect. 1.4). An analogous issue arises with the use of summed model weights in frequentist model averaging (Sect. 3.7).[4]

2.2.1 *Posterior Model Probabilities*

When calculating the posterior model probabilities, it can be difficult to determine the marginal likelihood in (2.3). In some settings, such as GLMs with conjugate priors for the parameters, the marginal likelihood can be expressed analytically [98], but in general we need to use an approximation. The most well-known approximation, which does not require specification of the priors for the parameters, is

$$p\left(y|m\right) \approx \exp\left(-\mathrm{BIC}_m/2\right) \tag{2.7}$$

where

$$\mathrm{BIC}_m = -2\log p\left(y|\widehat{\beta}_m, m\right) + p_m \log n, \tag{2.8}$$

and $\widehat{\beta}_m$ is the maximum likelihood estimate of β_m. The expression in (2.8) is the Bayesian information criterion for model m [91, 98, 114, 157].[5] The first term in (2.8) is influenced by the fit of the model, while the second can be thought of as a correction for overfitting which penalises more complex models.

Use of (2.7) in (2.2) leads to the approximation

$$p\left(m|y\right) \propto p\left(m\right)\exp\left(-\mathrm{BIC}_m/2\right), \tag{2.9}$$

which is sometimes referred to as the generalised BIC weight [114]. When we calculate the expression on the right-hand side of (2.9), BIC_m is often replaced by

$$\mathrm{BIC}_m - \min_k \mathrm{BIC}_k,$$

in order to avoid large arguments in the exponential function [32].[6]

Care is needed in specifying the value of n in (2.8). For a normal linear model, it is simply the number of observations. For a binomial model it is the total number

---

[4]We consider two contexts in which PIPs might be useful in Sect. 2.6.

[5]An alternative form is

$$\mathrm{BIC}_m = \log p\left(y|\widehat{\beta}_m, m\right) - \frac{1}{2}p_m \log n,$$

which [98] refer to as the Schwarz criterion [157]; multiplication of this by -2 allows BIC to be expressed on a deviance-scale. The same multiplication is used for DIC, WAIC (Sect. 2.3.1) and AIC (Sect. 3.2.1).

[6]Throughout the book, whenever a model weight based on an information criterion takes this form, it is implicit that calculation of the weight is carried out after this rescaling.

of Bernoulli trials. When using a log-linear model to analyse a contingency table it is the sum of the counts rather than the number of cells in the table [98, 145]. In the context of survival analysis, [171] suggested setting n equal to the number of uncensored observations. For a hierarchical model, the choice of n depends on the focus of the analysis [37, 98, 138, 139, 188]. For modelling survey data, [117] proposed a version of BIC that takes into account the design effect.

A higher-order approximation to the posterior model probabilities requires the priors for the parameters and their observed Fisher information matrix [96, 98, 116, 183]. It is similar in spirit to TIC, Takeuchi's information criterion, which has been suggested as an alternative to AIC in the frequentist setting [32] (Sect. 3.2.1).

Other approaches to approximating the posterior model probabilities have been proposed, involving marginal likelihoods [26, 28, 40, 41, 121, 132, 151] or Markov chain Monte Carlo (MCMC) methods [5, 9, 19, 23, 24, 29, 34, 37, 50, 71, 74, 75, 79, 82, 121, 142, 147, 179]. One conceptually-appealing method is reversible-jump MCMC (RJMCMC), in which we sample the parameter-space and model-space simultaneously [80]. However, RJMCMC can be prone to performance issues and be challenging to implement [8, 9, 37], to the extent that [83] recommend use of the approximation in (2.9). Recently, [8] have developed an approach which has the advantage of using the MCMC output obtained from fitting each model separately, and which exploits the relationships between parameters from different models [151].

2.2.2 *Choice of Priors*

Priors for models
A natural and common choice for the prior model probabilities is the uniform prior, in which $p(m) = 1/M$. The approximation to the posterior model probability in (2.9) then simplifies to the well-known BIC weight, given by

$$p\left(m\middle|y\right) \propto \exp\left(-\text{BIC}_m/2\right). \tag{2.10}$$

However, use of the uniform model-prior can have hidden implications. For example, if some of the possible predictor variables are highly correlated, we may have model-redundancy, in that some models will provide very similar estimates of θ. Use of a uniform prior will then dilute the prior probability allocated to any model which is not similar to the others [18, 57, 66, 76]. A method for dealing with this problem was proposed by [170], who suggested specifying prior model probabilities using the concept of the worth of a model, which is based on quantifying what we would expect to lose if we removed it from the model set when it is the true model. Another alternative is use of a Bernoulli prior in which each predictor variable has the same probability p of being included, independently of the others. The uniform prior is a special case, with $p = 0.5$, and therefore corresponds to a prior expectation that half the predictor variables will be included [37]. In order to have a less informative prior on model size, we might use a beta-prior for p [37, 105, 158].

Other approaches to specifying the model-prior involve empirical Bayes [37]; allowance for predictor variables being related, such as when some of the models include interaction terms [30]; and use of lower weights for models that are similar to others [66].

In order to allow for the possibility that the choice of model-prior may affect the form of the model-averaged posterior, [43] proposed use of credal model averaging, in which more than one model-prior is considered. This effectively allows one to preform a sensitivity analysis, in order to assess the extent to which the model-averaged posterior is influenced by the choice of model-prior. Further examples of the use of credal model averaging can be found in [44–46, 185].

Priors for model parameters

The posterior model probabilities can be sensitive to the choice of prior distribution for the parameters in a model, even if this prior would be regarded as non-informative in the single-model setting [8, 35, 93]. In particular, as mentioned in Sect. 2.2, the use of improper priors can lead to the Bayes factors, and hence the posterior model probabilities, not being well defined [10, 82, 93, 161]. It is also possible for apparently sensible priors for the parameters to cause the models to have conflicting implicit prior distributions for θ [85].

In the normal linear model setting, Zellner's g-prior has been used extensively, as it has several desirable properties, including computational convenience [37, 154]. It involves centring the predictor variables at zero, in order to remove any dependence between the intercept and the regression coefficients, and then specifying a joint prior for the intercept and error variance, plus a joint prior for the regression coefficients given the error variance. For model m, this leads to

$$p\left(\beta_{m0}, \sigma_m^2 \middle| m\right) \propto 1/\sigma_m^2$$

and a multivariate normal prior for the regression coefficients, with mean zero and covariance matrix

$$g_m \sigma_m^2 \left(X_m^\top X_m\right)^{-1},$$

where β_{m0}, σ_m^2, and X_m are the intercept, error variance and design matrix for model m respectively, and g_m is a hyperparameter [37, 187]. This prior has a nice interpretation, as it can be thought of as containing $1/g_m$ as much information as that in the data. The resulting posterior model probability is given by

$$p\left(m \middle| y\right) \propto p\left(m\right) \exp\left(-\mathrm{IC}_m/2\right),$$

where

$$\mathrm{IC}_m = -2 \log p\left(y \middle| \widehat{\beta}_m, m\right) + p_m \log g_m. \tag{2.11}$$

IC_m can be thought of as a generalised information criterion in which the correction for overfitting is $p_m \log g_m$. Setting g_m to be arbitrarily large, in order for the prior to be non-informative, can lead to strongly favouring the null model, an example of the

Lindley-Jeffreys paradox [37, 91, 98, 113]. Using $g_m = n$ gives the unit-information prior, which contains the same amount of information as a single observation, and leads to the posterior model probability being the generalised BIC weight in (2.9). Together with a uniform model-prior, this corresponds to using the BIC weight in (2.10), which was found perform well in a simulation study reported by [58]. An empirical Bayes procedure is also possible, in which the choice of g_m depends on the data [37]. As with the parameter p in the Bernoulli model-prior, we might also want to put a prior on g_m, rather than specify its value [105, 109, 111, 162, 186]. A version of the g-prior for high-dimensional normal linear models was proposed by [120].

For GLMs, [146] considered several approximations to the Bayes factors, including one that leads to the generalised BIC weight in (2.9). Extensions of Zellner's g-prior to this setting, including use of a prior on g_m, have been suggested by several authors; see [154] and the references therein. A calibrated information criterion (CIC) prior was proposed by [35]. This is based on the Jeffreys prior used in the single-model setting [91], and for model m it is given by

$$p\left(\beta_m \middle| m\right) = (2\pi)^{-p_m/2} \left|c_m^{-1} J\right|^{1/2},$$

where J is the observed Fisher information matrix for β_m and c_m is a hyperparameter. In conjunction with a uniform model-prior, this leads to the model-averaged posterior for θ being approximated by a multivariate normal distribution with mean $\widehat{\beta}_m$ and covariance matrix J^{-1}. In addition, the posterior probability for model m is approximated by

$$p\left(m \middle| y\right) \propto \exp\left(-\mathrm{CIC}_m/2\right), \tag{2.12}$$

where

$$\mathrm{CIC}_m = -2 \log p\left(y \middle| \widehat{\beta}_m, m\right) + p_m \log c_m,$$

which has the same form as (2.11). The right-hand side of (2.12) is known as the CIC weight; the BIC weight in (2.10) is a special case, corresponding to $c_m = n$.

2.3 Prediction-Based BMA

As mentioned in Sect. 2.1, classical BMA focusses attention on identification of the true model. Recently, several authors have considered use of prediction-based BMA [39, 102, 181]. In addition to being a more natural approach to model averaging, this has the distinct advantages of not requiring a prior for the models, being less sensitive to the priors for the parameters, and only requiring the usual MCMC output for each individual model.

There are currently two types of prediction-based BMA. The first involves a criterion based on a measure of the within-sample prediction error plus a correction term which allows for overfitting. The second uses cross validation and is therefore based on the error associated with prediction of observation i having fitted the model to all the data except that observation ($i = 1, \ldots, n$). The only difference between these approaches and classical BMA is that we combine posterior distributions using model weights that are not posterior model probabilities.

2.3.1 DIC and WAIC

In Bayesian model selection, the deviance information criterion (DIC) has long been used as an alternative to BIC [159, 160]. For model averaging, [15] suggested use of DIC weights, with BIC_m in (2.10) being replaced by

$$\text{DIC}_m = -2 \log p \left(y | \widehat{\beta}_m, m \right) + 2 p_m^{DIC}, \tag{2.13}$$

where $\widehat{\beta}_m$ is a point estimate of β_m and p_m^{DIC} is a correction for overfitting, often referred to as the effective number of parameters [159]. Common choices are

$$\widehat{\beta}_m = \text{E} \left(\beta_m | y, m \right) \tag{2.14}$$

and

$$p_m^{DIC} = 2 \, \text{var} \left\{ \log p \left(y | \beta_m \right) \right\} . \tag{2.15}$$

The posterior mean in (2.14) and posterior variance in (2.15) are estimated by the mean of the posterior MCMC sample for β_m and the variance of the posterior MCMC sample for $\log p \left(y | \beta_m \right)$ respectively. An alternative choice for p_m^{DIC} is possible [22, 73, 159], but this has the disadvantage of sometimes being negative.

DIC has much in common with AIC, which is also a prediction-based criterion (Sect. 3.2.1 and [32]). DIC model weights have been used in a range of applications, including ecology [60, 62, 115, 127, 167], fisheries [92, 177], medicine [143] and physics [112].

The other prediction-based measure we consider is the Watanabe-Akaike Information Criterion (WAIC) [72, 73, 89, 169, 174].[7] This is more Bayesian than DIC (and BIC), in that it replaces $\widehat{\beta}_m$ by the posterior distribution for β_m, and can work well in situations where DIC has problems [22]. The point estimate $\widehat{\beta}_m$ in DIC leads to underestimation of the prediction uncertainty, and hence to the possibility that use of DIC will lead to overfitting.[8] WAIC is also specified in terms of the pointwise predictive densities $p \left(y_i | \beta_m \right)$, rather than the joint predictive density $p \left(y | \beta_m \right)$, as the former has a close connection with cross validation [72] (Sect. 2.3.2). If the y_i are independent given the parameters, use of the joint density is equivalent to the pointwise-approach.

---

[7] See [72, 175] for the related Watanabe-Bayes Information Criterion (WBIC).

[8] See [160] for discussion of a modification to p_D that tries to compensate for this problem.

The value of WAIC for model m is given by

$$\text{WAIC}_m = -2 \sum_{i=1}^{n} \log p\left(y_i | y, m\right) + 2 p_m^{WAIC}, \tag{2.16}$$

where p_m^{WAIC} is again a correction for overfitting. The posterior predictive density in (2.16) is given by

$$p\left(y_i | y, m\right) = \int p\left(y_i | \beta_m, y, m\right) p\left(\beta_m | y, m\right) d\beta_m = \text{E}\left\{p\left(y_i | \beta_m, y, m\right)\right\}, \tag{2.17}$$

One choice for the correction term is

$$p_m^{WAIC} = \sum_{i=1}^{n} \text{var}\left\{\log p\left(y_i | \beta_m, y, m\right)\right\}. \tag{2.18}$$

As with DIC, the posterior mean in (2.17) and the posterior variance in (2.18) can be estimated by the mean of the posterior MCMC sample for $p\left(y_i | \beta_m, y, m\right)$ and the variance of the posterior MCMC sample for $\log p\left(y_i | \beta_m, y, m\right)$. As with DIC, an alternative choice for p_m^{WAIC} is possible [72, 73]; we consider that in (2.18) as it is closely related to leave-one-out cross validation (Sect. 2.3.2).

WAIC weights can be calculated using (2.10), with BIC_m replaced by WAIC_m. As DIC and WAIC are focussed on prediction, we would expect weights based on these criteria to be preferable to BIC weights, which are more focussed on identification of a true model [72]. As WAIC is more Bayesian than DIC, WAIC weights are based on a more reliable assessment of the prediction-uncertainty associated with each model. WAIC is also invariant to transformation of the parameters, whereas DIC will not be if we use (2.13), as the posterior mean is not transformation-invariant [159].[9] In addition, use of a pointwise-approach means that p_m^{WAIC} will be more stable than p_m^{DIC} [73].

As with BIC, when assessing the fit of a hierarchical model the exact form of DIC and WAIC will depend upon the focus of the analysis, as this will determine what we mean by prediction of a new observation [72, 122]; a similar issues arises when using AIC in the frequentist setting (Sect. 3.6.4).

2.3.2 Bayesian Stacking

Stacking is a cross-validation-based approach to model averaging that has a long history in the frequentist setting [164] (Sect. 3.2.3). Like the frequentist version, Bayesian stacking [181] uses a measure of out-of-sample prediction error, which

---

[9]This problem disappears if we use the posterior median [159] or posterior mode [22].

does not require a correction for overfitting. If a logarithmic scoring rule is used to summarise the prediction performance [78, 137], the model weights are chosen to be those that maximise the function

$$\sum_{i=1}^{n} \log \sum_{m=1}^{M} w_m p\left(y_i \middle| y_{-i}, m\right), \tag{2.19}$$

where w_m is the weight associated with model m and y_{-i} is the response vector y with y_i removed.[10] In order to maximise (2.19) using weights that lie on the unit simplex, we can use a constrained-optimisation method, such as quadratic programming [83]. Following [181], we refer to this approach as Bayesian stacking of predictive distributions (BSP).[11] Analogous to the form of the posterior predictive density used to calculate WAIC$_m$ in (2.16) (Sect. 2.3.1), we have

$$p\left(y_i \middle| y_{-i}, m\right) = \int p\left(y_i \middle| \beta_m, y_{-i}, m\right) p\left(\beta_m \middle| y_{-i}, m\right) d\beta_m = \mathrm{E}\left\{p\left(y_i \middle| \beta_m, y_{-i}, m\right)\right\},$$

where the posterior mean on the right-hand side is now with respect to $p\left(\beta_m \middle| y_{-i}, m\right)$, and can be estimated by the mean of the corresponding posterior MCMC sample for $p\left(y_i \middle| \beta_m, y_{-i}, m\right)$.

As computational effort will often be an important consideration in the Bayesian setting, [181] proposed use of Pareto-smoothed importance sampling [168], which only requires a single fit to the data for each model. On the other hand, if the sample size is small estimation of the weights may be unstable [181], an example of which arises in the toxicity example (Sect. 2.4.2).

Determining posterior model weights by minimising an objective function has also been suggested by [81, 172]. Likewise, in the context of forecasting in economic time series, [59, 63] have proposed using an estimate of out-of-sample prediction error to determine model weights. A decision-theoretic approach to BMA, also based on prediction error, was used by [16] in the context of high-dimensional multivariate regression models.

When n is large, BSP might be expected to produce weights that are similar to those based on WAIC, as the latter is asymptotically equivalent to use of Bayesian leave-one-out cross validation for model selection [174]. A discussion of the relative merits of DIC, WAIC and Bayesian cross validation can be found in [72].

In related work, interpretation of a model-averaged posterior as a mixture distribution has been advocated by [94]; see also [181].[12] As with the approach of [172], this leads to improper priors for the model parameters being acceptable.

---

[10]Using a sample of size $n - 1$ to estimate the prediction error associated with a model that is fitted to n observations can lead to overestimation of the prediction uncertainty, and a bias-adjustment can be made to allow for this [72].

[11]Alternative versions of prediction-based BMA are discussed in detail by [181].

[12]Use of BMA for averaging mixture models has been considered by [176].

Use of BSP can be motivated by the fact that classical BMA has been shown to have poorer prediction performance than frequentist stacking, particularly when the true model is not in the model set [33, 53, 178, 181]. Use of classical BMA leads to an asymptotic weight of one for the model closest to the true data-generating mechanism (in terms of Kullback-Leibler divergence). In contrast, BSP finds the optimal combination of predictive distributions that is closest to the data-generating mechanism (in terms of the scoring rule), and the asymptotic BSP weights can all be less than one [181]. A similar motivation led to the idea of Bayesian model combination in the machine-learning literature [100, 124, 126]. If one of the models is a good approximation to the data-generating mechanism, BSP may not perform as well as classical BMA when n is small.

2.4 Examples

2.4.1 Ecklonia Density

For the ecklonia example in Sect. 1.3.2, the model weights obtained using BIC, DIC and WAIC are shown in Table 2.1, together with those for AIC (Sect. 1.3.2). The most striking difference is the greater weight given to model 1 by BIC, compared to the other methods. This might be expected, as the DIC and WAIC weights are prediction-based, and are therefore similar to the AIC weights. The estimates of the effective number of parameters associated with DIC and WAIC are also shown in Table 2.1. These are lower for WAIC, as it uses a better measure of prediction uncertainty, which requires less of a correction for overfitting [72] (Sect. 2.3.1).

Figure 2.1 shows each of the model-averaged posterior distributions for the mean density of ecklonia in each zone, together with the posterior for each model.[13] Note the bi-modality of the model-averaged posteriors for zone 1, a feature that is not uncommon in BMA. The posterior means and 95% credible intervals are given in Table 2.2, together with the corresponding model-averaged estimates and 95% confidence intervals based on AIC weights (Table 1.4). The main difference is between BIC and the other methods; it generally provides a narrower credible interval, as it gives most weight to model 1. For zone 2, the posterior means from the two models are similar, which leads to little difference between the model-averaged posterior means.

---

[13] We used a uniform prior between 0 and 10^3 for $1/k$ and a $N(0, 10^4)$ prior for each $\log \mu$. For DIC, we used $\widehat{\beta}_m = E\left(\beta_m \mid y, m\right)$, with $\beta_1 = (\log \mu, 1/k)^\top$ and $\beta_2 = (\log \mu_1, \log \mu_2, \log \mu_3, 1/k)^\top$.

Table 2.1 Model weights for the ecklonia study, based on AIC, BIC, DIC and WAIC, together with the effective number of parameters, as estimated using DIC and WAIC, and the true number of parameters

Model (m)	AIC	BIC	DIC	WAIC	p_m	p_m^{DIC}	p_m^{WAIC}
1	0.276	0.840	0.367	0.201	2	2.1	1.7
2	0.724	0.160	0.633	0.799	4	4.5	2.7

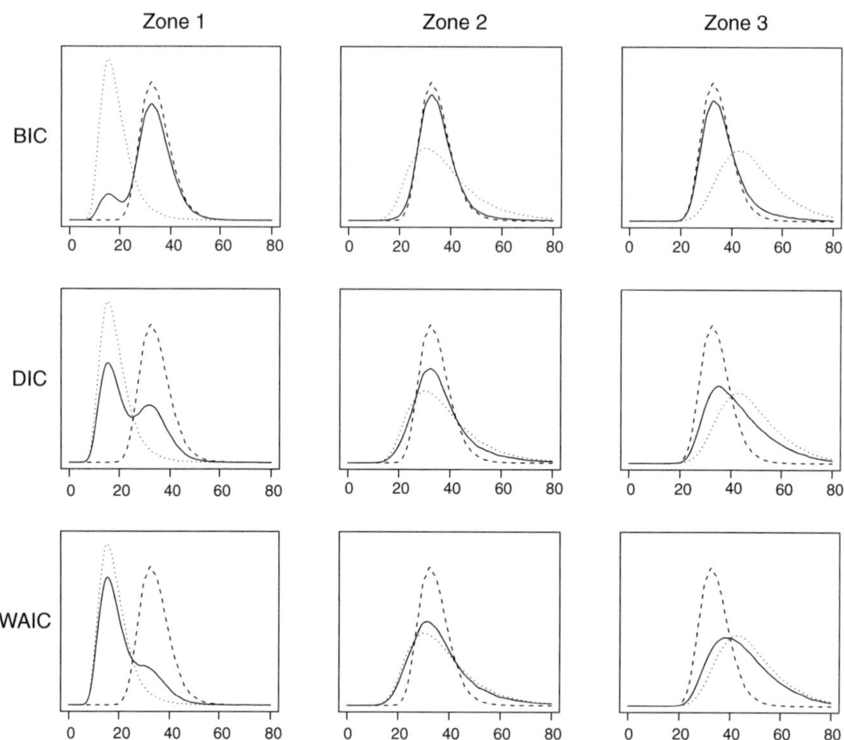

Fig. 2.1 Model-averaged posterior distributions for mean ecklonia density in each of three zones (solid), together with the posterior distributions for model 1 (dashed) and model 2 (dotted), for three types of model weight (BIC, DIC and WAIC). Density is given as individuals per quadrat

2.4.2 *Toxicity of a Pesticide*

For the pesticide example in Sect. 1.3.4, the quantities of interest were the doses that lead to 50% or 90% of individuals being affected, separately for each sex. The BSP weights will depend upon whether we remove a group of moths or an individual moth, corresponding to assessing the models on slightly different prediction problems, analogous to the choice of focus in hierarchical models (Sect. 2.3.1). Likewise,

Table 2.2 Posterior means and 95% credible intervals for the mean density (individuals per quadrat) of ecklonia in three zones, for each model and using Bayesian model averaging based on BIC, DIC or WAIC weights. The corresponding estimates and 95% confidence intervals from frequentist model averaging using AIC weights are also shown for comparison

		Mean	Lower	Upper
Zone 1	Model 1	34.5	24.6	48.4
	Model 2	18.6	10.0	34.1
	AIC	20.4	9.9	42.0
	BIC	32.0	13.0	47.8
	DIC	24.5	10.6	44.7
	WAIC	21.8	10.3	42.6
Zone 2	Model 1	34.5	24.6	48.4
	Model 2	38.0	19.1	75.0
	AIC	33.6	18.5	61.1
	BIC	35.1	23.2	53.8
	DIC	36.8	20.2	68.4
	WAIC	37.3	19.6	71.6
Zone 3	Model 1	34.5	24.6	48.4
	Model 2	48.4	29.1	79.8
	AIC	41.7	25.9	70.7
	BIC	36.7	24.8	60.9
	DIC	43.3	26.0	74.9
	WAIC	45.6	26.9	77.3

treating the data as binomial or binary will affect the WAIC weights. Interestingly, the same cannot be said of DIC, with both terms in (2.13) being identical for the two model-formulations, reflecting another problem associated with this criterion.

We consider both the binomial and binary model-formulations when determining the weights (the choice of formulation does not affect the posterior distribution for each parameter). As discussed in Sect. 2.2.1, the derivation of BIC leads to using $n = 240$ when calculating the BIC weight, corresponding to the binary-model formulation. For completeness, we also include the binomial model formulation when calculating BIC weights. For the binomial model, we have $n = 12$ and use leave-one-group-out cross validation for BSP, while for the binary model we have $n = 240$ and use leave-one-individual-out cross validation for BSP.[14]

The model weights based on BIC, DIC, WAIC and BSP are shown in Table 2.3, together with the AIC weights (Sect. 1.3.4). In all but one case, model 3 has the most weight, while model 1 has negligible weight in all cases. The largest differences are between BSP and the other methods. Using the binomial-model formulation, model

---

[14]We used a N(0, 10^4) prior for each intercept and slope. For DIC, we used $\widehat{\beta}_m = \mathrm{E}\left(\beta_m | y, m\right)$, with β_m being the vector of intercepts and slopes for model m.

Table 2.3 Model weights for the toxicity experiment obtained using BIC, DIC, WAIC and BSP, using two model-formulations. Frequentist AIC weights, which do not depend on the model-formulation, are also shown for comparison

Model	AIC	Binomial model				Binary model			
		BIC	DIC	WAIC	BSP	BIC	DIC	WAIC	BSP
1	0.004	0.005	0.005	0.001	0.000	0.026	0.005	0.004	0.031
2	0.235	0.246	0.217	0.189	0.000	0.277	0.217	0.238	0.488
3	0.553	0.578	0.577	0.587	1.000	0.653	0.577	0.568	0.120
4	0.209	0.171	0.201	0.222	0.000	0.043	0.201	0.190	0.361

Table 2.4 Posterior means and 95% credible intervals for the dose-levels (μg) of trans-cypermethrin that leads to 50% or 90% of individuals being affected, for each of four models, separately for each sex

Probability affected	Model	Male			Female		
		Mean	Lower	Upper	Mean	Lower	Upper
0.5	1	6.8	5.4	8.4	6.8	5.4	8.4
	2	4.8	3.5	6.4	9.7	7.1	13.2
	3	4.8	3.7	6.2	10.1	7.1	14.5
	4	4.8	3.5	6.2	10.1	7.1	14.6
0.9	1	31.2	20.5	49.6	31.2	20.5	49.6
	2	19.8	12.8	31.8	40.9	25.3	69.6
	3	15.7	10.2	24.9	59.3	29.7	124.5
	4	16.3	10.1	27.3	58.3	27.8	141.7

3 is given all the weight by BSP, but estimation of this weight was unreliable, with several local optima being encountered en route. For the binary-model case, model 2 has the highest BSP weight, and the weight for model 4 is larger than that for model 3. In general, we would expect the WAIC weights based on the binary-model formulation to be more reliable than those from the binomial-model formulation, as the summations used to calculate WAIC are based on more terms. BIC is the only method that gives negligible weight to model 4.

Table 2.4 shows the posterior mean and 95% credible interval obtained from each model, for each of the four choices of θ. These are similar to the frequentist estimates and 95% confidence intervals given in Sect. 1.3.4, the main difference being the amount of uncertainty associated with the dose that leads to 90% of females being affected; the upper credible limits for models 3 and 4 are substantially higher than the corresponding upper confidence limits in Table 1.10.

The model-averaged posterior means and 95% credible intervals are shown in Table 2.5. For simplicity we only consider the binary-model formulation, as the BIC weights correctly use $n = 240$ and the BSP weights are estimated more reliably. For all methods, the results are broadly similar to the model-averaged estimates and 95% confidence intervals based on AIC weights (Table 1.10). The main difference

Table 2.5 Model-averaged posterior means and 95% credible intervals for the dose-levels (μg) of trans-cypermethrin that leads to 50% or 90% of individuals being affected, using weights based on BIC, DIC, WAIC and BSP, separately for each sex, using a binary-model formulation

Probability affected		Male			Female		
		Mean	Lower	Upper	Mean	Lower	Upper
0.5	BIC	4.8	3.6	6.6	9.9	6.7	14.1
	DIC	4.8	3.6	6.3	10.0	6.9	14.2
	WAIC	4.8	3.6	6.3	10.0	7.0	14.2
	BSP	4.8	3.5	6.7	9.8	6.7	13.9
0.9	BIC	17.3	10.4	30.6	53.3	26.8	116.4
	DIC	16.8	10.3	28.4	55.1	27.3	121.8
	WAIC	16.8	10.4	28.3	54.6	27.2	118.4
	BSP	18.4	10.8	32.2	49.0	25.7	111.0

is again for the dose that leads to 90% of females being affected, with the credible limits being slightly higher than the corresponding confidence limits.

2.5 Discussion

BMA has many appealing aspects, including conceptual simplicity, a natural ability to allow for uncertainty, and the use of a posterior distribution to represent uncertainty. The latter will often be more informative than a frequentist point estimate and confidence interval, as shown by the ecklonia example (Fig. 2.1). As in the single-model setting, BMA is also transformation-invariant, in that the scale on which we perform BMA does not matter, unlike some procedures in frequentist model averaging (Sects. 3.2 and 3.6).[15]

As discussed in Chap. 1, model averaging is useful when θ has the same interpretation in all models. When we are interested in estimating a regression coefficient and the interpretation of this coefficient is the same in all models (Sect. 1.4), the model-averaged posterior distribution will often contain a spike at zero, and this may be regarded as an advantage [87] or a disadvantage [57].

There are several issues associated with classical BMA:

1. By definition it allows for the possibility that one of the smaller models is true, which we regard as less natural than assuming that the largest model is true or that the true model is not in the set;
2. Model averaging is concerned with estimation rather than identification of a true model (Sect. 1.4);

---

[15] As noted in Sect. 2.3.1. the exception occurs when we use $\widehat{\beta}_m = \mathrm{E}\left(\beta_m \mid y, m\right)$ in (2.13), as the DIC weight then depends on the parametrisation.

3. If θ is the expected value of Y for specific values of the predictor variables, classical BMA may not have good frequentist properties unless the priors depend on n [180];
4. The posterior model probabilities can be sensitive to the choice of priors for the parameters, and the implicit prior for θ can vary between models [32].

By definition, classical BMA will have perfect frequentist properties when we can assume that the complete model, including the priors for the models and the parameters, is true [37, 47, 85, 148, 180]. Thus we need to assume that the data are generated in the following three stages:

1. A model is selected at random from the set of candidate models, using the prior model probabilities.
2. The parameter values for this model are generated using the relevant prior distributions.
3. The data are generated from the selected model and parameter values.

Much of the literature that discusses optimality-properties of classical BMA, including prediction-based properties, does so under the implicit assumption that this complete model is true [148].

Prediction-based BMA provides a promising means of overcoming some of the difficulties associated with classical BMA, as it is focussed on estimation rather than identification of the true model, and allows us to avoid assuming that one of the models is true. It would be interesting to assess the sampling properties of BSP, compared to frequentist stacking (Sect. 3.2.3). In addition, it would be useful to have a variation of BSP that is focussed on obtaining an optimal model-averaged posterior credible interval, using an interval-focussed scoring rule [182]. A discussion of the relative merits of DIC, WAIC and cross validation in the context of model selection can be found in [72, 73].

2.6 Related Literature

A review of classical BMA is available in [65]. Much of the literature in this area has involved discussion of the computational challenges [5, 24, 27, 31, 37, 82, 90]. In the context of high-dimensional linear models, use of non-local priors for the parameters has been advocated by [152], as these can lead to asymptotically quicker removal of spurious predictor variables. In related work on high-dimensional normal linear models, [20] have shown that a model-prior which has a sufficiently large correction for overfitting can work well at removing such predictors.

Several methods have been proposed for efficiently searching the model-space in classical BMA, including adaptive sampling, evolutionary Monte Carlo, the leaps-and-bounds algorithm, MC^3, multi-set model selection, Occam's window, random searching, and stochastic-search variable selection [14, 38, 61, 75, 86, 103, 110, 118, 119, 133, 145]. In practice, use of RJMCMC will often lead to considering only a fraction of the possible models [101, 130]. An iterative procedure was used by [6], in the context of analysing high-dimensional microarray data, where the number of potential models can be extremely large. A useful review of methods for variable selection in the Bayesian setting is provided by [136].

In model selection, the sensitivity of Bayes factors to the prior distributions for the parameters has led to new types of Bayes factor being proposed, some of which have links to information criteria [1, 10, 67, 69, 70, 134, 135, 140].

In the normal linear model setting, [7] showed that the median-probability-model, which contains all predictors with a PIP of at least 0.5, is the optimal choice of a best model when we use a squared-error loss function. Interestingly, this model may not be the one with the highest posterior probability [56]. Another setting in which the PIP might be useful is the analysis of microarray data, when we wish to predict the diagnostic category of a tissue sample from its expression array phenotype [184], the aim being to determine a (hopefully small) set of genes that can be used in a diagnostic test. In this context, there is a natural balance in the model set, and the PIP for a gene provides an index of its relevance to the test.

Some authors have advocated consideration of the joint PIP for each pair of predictor variables, the idea being to assess their potential joint impact on the response variable [48, 54, 55, 77, 88, 104, 106, 166]. As with the marginal PIP, such a measure will be influenced by the choice of model set, and a more useful summary might be obtained by comparing model-averaged posterior distributions for the expectation of the response variable at suitably-chosen values of the two predictor variables of interest (Sect. 1.4).[16]

Use of BIC weights to average over frequentist estimates has been considered by [128, 156]. BIC is also related to the concept of minimum description length (MDL) in communication theory [32, 83, 150].

An extensive review of objective priors, for both models and parameters, has been provided by [42], who argue that we might want the choice of prior for the parameters in each model to depend on whether our focus is on determining the posterior model probabilities (as in classical BMA) or on the posterior distribution for θ. In the single-model setting, work on probability-matching priors has been motivated by a desire to make Bayesian methods reliable from a frequentist perspective [49], but they have yet to be applied in BMA.

As in the frequentist setting, several Bayesian approaches to combining forecasts have been developed, especially for economic time series [3, 17, 51, 141, 163], but also in other areas [99, 108]. There has also been some comparison of Bayesian and frequentist approaches to combining forecasts [95, 123]. Excellent reviews of the use of BMA in economics are also available [129, 161, 162]. The following BMA packages are available in R [144]:

1. `loo` provides WAIC and BSP model weights
2. `BayesianTools` calculates DIC, WAIC and Bayes factors

---

[16]This is distinct from inspection of the model-averaged joint posterior distribution of two model parameters, which can be checked in the usual way for posterior dependence between parameters.

3. `rjmcmc` implements the methods developed by [8]
4. `glmBfp` and `hypergsplines` are for generalised additive models [153, 155]
5. `BDgraph` is relevant to graphical models [125]
6. `eDMA` provides Bayesian dynamic model averaging [149]
7. `madr` applies BMA to causal inference [21, 173, 189]
8. `BMS`, `BAS` and `BMA` are reviewed by [2]
9. `BayesFactor`, `BayesVarSel` and `mombf` are reviewed by [64]

A review of other BMA software and online resources can be found in [162].

References

1. Aitkin, M.: Posterior Bayes factors. J. Roy. Stat. Soc. B. Methodol. **53**, 111–142 (1991)
2. Amini, S.M., Parmeter, C.F.: Bayesian model averaging in R. J. Econ. Soc. Meas. **36**, 253–287 (2011)
3. Anandalingam, G., Chen, L.: Linear combination of forecasts: a general Bayesian model. J. Forecasting **8**, 199–214 (1989)
4. Ando, T., Tsay, R.: Predictive likelihood for Bayesian model selection and averaging. Int. J. Forecasting **26**, 744–763 (2010)
5. Andrieu, C., Doucet, A., Robert, C.P.: Computational advances for and from Bayesian analysis. Stat. Sci. **19**, 118–127 (2004)
6. Annest, A., Bumgarner, R.E., Raftery, A.E., Yeung, K.Y.: Iterative Bayesian model averaging: a method for the application of survival analysis to high-dimensional microarray data. BMC Bioinform. **10**, 72 (2009)
7. Barbieri, M.M., Berger, J.O.: Optimal predictive model selection. Ann. Stat. **32**, 870–897 (2004)
8. Barker, R.J., Link, W.A.: Bayesian multimodel inference by RJMCMC: a Gibbs sampling approach. Am. Stat. **67**, 150–156 (2013)
9. Bartolucci, F., Scaccia, L., Mira, A.: Efficient Bayes factor estimation from the reversible jump output. Biometrika **93**, 41–52 (2006)
10. Berger, J.O., Pericchi, L.R.: The intrinsic Bayes factor for model selection and prediction. J. Am. Stat. Assoc. **91**, 109–122 (1996)
11. Berger, J.O., Ghosh, J.K., Mukhopadhyay, N.: Approximations and consistency of Bayes factors as model dimension grows. J. Stat. Plan. Infer. **112**, 241–258 (2003)
12. Berger, J.O., Molina, G.: Posterior model probabilities via pathbased pairwise priors. Stat. Neerl. **59**, 3–15 (2005)
13. Bernardo, J.M., Smith, A.F.M.: Bayesian Theory. Wiley, New York (1994)
14. Bottolo, L., Richardson, S.: Evolutionary stochastic search for Bayesian model exploration. Bayesian Anal. **5**, 583–618 (2010)
15. Brooks, S.P.: Discussion of Spiegelhalter, D.J., Best, N.G., Carlin, B.P., Van Der Linde, A.: Bayesian measures of model complexity and fit. J. R. Stat. Soc. **64**, 616–618 (2002)
16. Brown, P.J., Vannucci, M., Fearn, T.: Bayes model averaging with selection of regressors. J. Roy. Stat. Soc. B Methodol. **64**, 519–536 (2002)
17. Bunn, D.W.A.: Bayesian approach to the linear combination of forecasts. Oper. Res. Quart. **26**, 325–329 (1975)
18. Burnham, K.P., Anderson, D.R.: Model Selection and Multimodel Inference: A Practical Information-Theoretic Approach, 2nd edn. Springer, New York (2002)
19. Carlin, B.P., Chib, S.: Bayesian model choice via Markov chain Monte Carlo methods. J. Roy. Stat. Soc. B. Methodol. **57**, 473–484 (1995)

20. Castillo, I., Schmidt-Hieber, J., Van der Vaart, A.: Bayesian linear regression with sparse priors. Ann. Stat. **43**, 1986–2018 (2015)
21. Cefalu, M., Dominici, F., Arvold, N., Parmigiani, G.: Model averaged double robust estimation. Biometrics **73**, 410–421 (2017)
22. Celeux, G., Forbes, F., Robert, C.P., Titterington, D.M.: Deviance information criteria for missing data models. Bayesian Anal. **1**, 651–673 (2006)
23. Chen, M.-H., Shao, Q.-M.: On Monte Carlo methods for estimating ratios of normalizing constants. Ann. Stat. **25**, 1563–1594 (1997)
24. Chen, M.-H., Shao, Q.-M., Ibrahim, J.G.: Monte Carlo Methods in Bayesian Computation. Springer, New York (2000)
25. Chen, M.-H., Ibrahim, J.G.: Conjugate priors for generalized linear models. Stat. Sinica. **13**, 461–476 (2003)
26. Chib, S.: Marginal likelihood from the Gibbs output. J. Am. Stat. Assoc. **90**, 1313–1321 (1995)
27. Chib, S.: Monte Carlo methods and Bayesian computation: overview. In: Smelser, N.J., Baltes, P.B. (eds.) International Encyclopedia of the Social and Behavioral Sciences: Statistics. Elsevier Science, Oxford (2001)
28. Chickering, D.M., Heckerman, D.: Efficient approximations for the marginal likelihood of Bayesian networks with hidden variables. Mach. Learn. **29**, 181–212 (1997)
29. Ching, J., Chen, Y.-C.: Transitional Markov chain Monte Carlo method for Bayesian model updating, model class selection, and model averaging. J. Eng. Mech. **133**, 816–832 (2007)
30. Chipman, H.: Bayesian variable selection with related predictors. Can. J. Stat. **24**, 17–36 (1996)
31. Chipman, H., George, E.I., McCulloch, M., Clyde, D.P.F., Stine, R.A.: The practical implementation of Bayesian model selection. Inst. Math. S. **38**, 65–134 (2001)
32. Claeskens, G., Hjort, N.L.: Model Selection and Model Averaging, vol. 330. Cambridge University Press, Cambridge (2008)
33. Clarke, B.: Comparing Bayes model averaging and stacking when model approximation error cannot be ignored. J. Mach. Learn. Res. **4**, 683–712 (2003)
34. Clyde, M., Desimone, H., Parmigiani, G.L.: Prediction via orthogonalized model mixing. J. Am. Stat. Assoc. **91**, 1197–1208 (1996)
35. Clyde, M.: Model uncertainty and health effect studies for particulate matter. Environmetrics **11**, 745–763 (2000)
36. Clyde, M.: Model averaging. In: Press, S.J. (ed.) Subjective and Objective Bayesian Statistics, 2nd edn. Wiley-Interscience, New Jersey (2003)
37. Clyde, M., George, E.I.: Model uncertainty. Stat. Sci. **19**, 81–94 (2004)
38. Clyde, M.A., Ghosh, J., Littman, M.L.: Bayesian adaptive sampling for variable selection and model averaging. J. Comput. Graph. Stat. **20**, 80–101 (2011)
39. Clyde, M., Iversen, E.S.: Bayesian model averaging in the M-open framework. In: Damien, P., Dellaportas, P., Polson, N.G., Stephens, D.A. (eds.) Bayesian Theory and Applications. Oxford University Press, Oxford (2013)
40. Congdon, P.: Bayesian model choice based on Monte Carlo estimates of posterior model probabilities. Comput. Stat. Data Anal. **50**, 346–357 (2006)
41. Congdon, P.: Model weights for model choice and averaging. Stat. Methodol. **4**, 143–157 (2007)
42. Consonni, G., Fouskakis, D., Liseo, B., Ntzoufras, I.: Prior distributions for objective Bayesian analysis. Bayesian Anal. **13**, 627–679 (2018)
43. Corani, G., Zaffalon, M.: Credal model averaging: an extension of Bayesian model averaging to imprecise probabilities. In: Joint European Conference on Machine Learning and Knowledge Discovery in Databases. Springer, Heidelberg (2008)
44. Corani, G., Antonucci, A.: Credal ensembles of classifiers. Comput. Stat. Data Anal. **71**, 818–831 (2014)
45. Corani, G., Mignatti, A.: Credal model averaging for classification: representing prior ignorance and expert opinions. Int. J. Approx. Reason. **56**, 264–277 (2015)

46. Corani, G., Mignatti, A.: Robust Bayesian model averaging for the analysis of presence-absence data. Environ. Ecol. Stat. **22**, 513–534 (2015)
47. Cox, D.R.: Principles of Statistical Inference. Cambridge University Press, Cambridge (2006)
48. Cuaresma, J.C., Grün, B., Hofmarcher, P., Humer, S., Moser, M.: Unveiling covariate inclusion structures in economic growth regressions using latent class analysis. Eur. Econ. Rev. **81**, 189–202 (2016)
49. Datta, G.S., Mukerjee, R.: Probability Matching Priors: Higher Order Asymptotics. Springer, New York (2004)
50. DiCiccio, T.J., Kass, R.E., Raftery, A., Wasserman, L.: Computing Bayes factors by combining simulation and asymptotic approximations. J. Am. Stat. Assoc. **92**, 903–915 (1997)
51. Diebold, F.X., Pauly, P.: The use of prior information in forecast combination. Int. J. Forecasting **6**, 503–508 (1990)
52. Domingos, P.: Why does bagging work? A Bayesian account and its implications. In: Proceedings of the Third International Conference on Knowledge Discovery and Data Mining, pp. 155–158 (1997)
53. Domingos, P.: Bayesian averaging of classifiers and the overfitting problem. In: Proceedings of the Seventeenth International Conference on Machine Learning, pp. 223–230 (2000)
54. Doppelhofer, G., Weeks, M.: Jointness of growth determinants. J. Appl. Econ. **24**, 209–244 (2009)
55. Doppelhofer, G., Weeks, M.: Jointness of growth determinants: reply to comments by Rodney Strachan, Eduardo Ley and Mark FJ Steel. J. Appl. Econ. **24**, 252–256 (2009)
56. Drachal, K.: Comparison between Bayesian and information-theoretic model averaging: fossil fuels prices example. Energ. Econ. **74**, 208–251 (2018)
57. Draper, D.: Model uncertainty yes, discrete model averaging maybe. Stat. Sci. **14**, 405–409 (1999)
58. Eicher, T.S., Papageorgiou, C., Raftery, A.E.: Default priors and predictive performance in BMA, with application to growth determinants. J. Appl. Econ. **26**, 30–55 (2011)
59. Eklund, J., Karlsson, S.: Forecast combination and model averaging using predictive measures. Econ. Rev. **26**, 329–363 (2007)
60. Ellison, A.M.: Bayesian inference in ecology. Ecol. Lett. **7**, 509–520 (2004)
61. Fan, T.-H., Wang, G.-T., Yu, J.-H.: A new algorithm in Bayesian model averaging in regression models. Commun. Stat. Simul. **43**, 315–328 (2014)
62. Farnsworth, M.L., Hoeting, J.A., Thompson Hobbs, N., Miller, M.W.: Linking chronic wasting disease to mule deer movement scales: a hierarchical Bayesian approach. Ecol. Appl. **16**, 1026–1036 (2006)
63. Feldkircher, M.: Forecast combination and BMA: a prior sensitivity analysis. J. Forecasting **31**, 361–376 (2012)
64. Forte, A., Garcia-Donato, G., Steel, M.F.J.: Methods and tools for Bayesian variable selection and model averaging in normal linear regression. Department of Statistics working paper, University of Warwick (2017)
65. Fragoso, T.M., Bertoli, W., Louzada, F.: Bayesian model averaging: a systematic review and conceptual classification. Int. Stat. Rev. (2017). https://doi.org/10.1111/insr.12243
66. Garthwaite, P.H., Mubwandarikwa, E.: Selection of weights for weighted model averaging. Aust. NZ. J. Stat. **52**, 363–382 (2010)
67. Geisser, S., Eddy, W.F.: A predictive approach to model selection. J. Am. Stat. Assoc. **74**, 153–160 (1979)
68. Gelfand, A.E., Dey, D.K., Chang, H.: Model determination using predictive distributions with implementation via sampling-based methods. Technical report 462. Department of Statistics, Stanford University (1992)
69. Gelfand, A., Dey, D.K.: Bayesian model choice: asymptotics and exact calculations. J. R. Stat. Soc. B. Methodol. **56**, 501–514 (1994)
70. Gelfand, A.E.: Model determination using sampling-based methods. In: Gilks, W.R., Richardson, S., Spiegelhalter, D.J. (eds.) Markov Chain Monte Carlo in Practice, pp. 145–162. Chapman and Hall (1996) In: Bernardo, J.M., Berger, J.O., Dawid, A.P., Smith, A.F.M. (eds.) Bayesian Statistics, vol. 6, pp. 175–177. Oxford University Press (1999)

71. Gelman, A., Meng, X.-L.: Simulating normalizing constants: from importance sampling to bridge sampling to path sampling. Stat. Sci. **13**, 163–185 (1998)
72. Gelman, A., Hwang, J., Vehtari, A.: Understanding predictive information criteria for Bayesian models. Stat. Comput. **24**, 997–1016 (2014)
73. Gelman, A., Carlin, J.B., Stern, H.S., Dunson, D.B., Vehtari, A., Rubin, D.B.: Bayesian Data Analysis. CRC Press, Boca Raton (2014)
74. George, E.I., McCulloch, R.E.: Variable selection via Gibbs sampling. J. Am. Stat. Assoc. **88**, 881–889 (1993)
75. George, E.I., McCulloch, R.E.: Approaches for Bayesian variable selection. Stat. Sin. **7**, 339–373 (1997)
76. George, E.I., Discussion of Clyde, M.A.: BMA and model search strategies. In: Bernardo, J.M., Berger, J.O., Dawid, A.P., Smith, A.F.M. (eds.) Bayesian Statistics, vol. 6, pp. 175–177. Oxford University Press (1999)
77. Ghosh, J., Ghattas, A.E.: Bayesian variable selection under collinearity. Am. Stat. **69**, 165–173 (2015)
78. Gneiting, T., Raftery, A.E.: Strictly proper scoring rules, prediction, and estimation. J. Am. Stat. Assoc. **102**, 359–378 (2007)
79. Godsill, S.J.: On the relationship between Markov chain Monte Carlo methods for model uncertainty. J. Comput. Graph. Stat. **10**, 230–248 (2001)
80. Green, P.J.: Reversible jump Markov chain Monte Carlo computation and Bayesian model determination. Biometrika **82**, 711–732 (1995)
81. Gutiérrez-Peña, E., Walker, S.G.: Statistical decision problems and Bayesian nonparametric methods. Int. Stat. Rev. **73**, 309–330 (2005)
82. Han, C., Carlin, B.P.: Markov chain Monte Carlo methods for computing Bayes factors. J. Am. Stat. Assoc. **96**, 1122–1132 (2001)
83. Hastie, T., Tibshirani, R., Friedman, J.: The Elements of Statistical Learning, vol. 2, no. 1. Springer, New York (2009)
84. Hernández, B., Raftery, A.E., Pennington, S.R., Parnell, A.C.: Bayesian additive regression trees using Bayesian model averaging. Stat. Comput. **28**, 869–890 (2018)
85. Hjort, N.L., Claeskens, G.: Frequentist model average estimators. J. Am. Stat. Assoc. **98**, 879–945 (2003)
86. Hoegh, A., Maiti, D., Leman, S.: Multiset model selection. J. Comput. Graph. Stat. (2018). https://doi.org/10.1080/10618600.2017.1379408
87. Hoeting, J.A., Madigan, D., Raftery, A.E., Volinsky, C.T.: Bayesian model averaging: a tutorial. Stat. Sci. **14**, 382–401 (1999)
88. Hofmarcher, P., Cuaresma, J.C., Grün, B., Humer, S., Moser, M.: Bivariate jointness measures in Bayesian model averaging: solving the conundrum. J. Macroecon. **57**, 150–165 (2018)
89. Hooten, M.B., Thompson Hobbs, N.: A guide to Bayesian model selection for ecologists. Ecol. Monogr. **85**, 3–28 (2015)
90. Hubin, A., Storvik, G.: Mode jumping MCMC for Bayesian variable selection in GLMM. Comput. Stat. Data Anal. **127**, 281–297 (2018)
91. Jeffreys, H.: Theory of Probability, 3rd edn. Oxford University Press, Oxford (1961)
92. Jiao, Y., Reid, K., Smith, E.: Model selection uncertainty and BMA in fisheries recruitment modeling. In: Beamish, R.J., Rothschild, B.J. (eds.) The Future of Fisheries Science in North America, pp. 505–524. Springer, Dordrecht (2009)
93. Kadane, J.B., Lazar, N.A.: Methods and criteria for model selection. J. Am. Stat. Assoc. **99**, 279–290 (2004)
94. Kamary, K., Mengersen, K., Robert, C.P., Rousseau, J.: Testing hypotheses via a mixture estimation model (2014). arXiv preprint: arXiv:1412.2044
95. Kapetanios, G., Labhard, V., Price, S.P.: Forecasting using Bayesian and information-theoretic model averaging: an application to UK inflation. J. Bus. Econ. Stat. **26**, 33–41 (2008)
96. Kashyap, R.L.: Optimal choice of AR and MA parts in autoregressive moving average models. IEEE Trans. Pattern Anal. **4**, 99–104 (1982)
97. Kass, R.E.: Bayes factors in practice. J. Roy. Stat. Soc. D Stat. **42**, 551–560 (1993)

98. Kass, R.E., Raftery, A.E.: Bayes factors. J. Am. Stat. Assoc. **90**, 773–795 (1995)
99. Kiartzis, S., Kehagias, A., Bakirtzis, A., Petridis, V.: Short term load forecasting using a Bayesian combination method. Int. J. Electr. Power **19**, 171–177 (1997)
100. Kim, H.-C., Ghahramani Z.: Bayesian classifier combination. In: Proceedings the 15th International Conference Artificial Intelligence and Statistics, pp. 619–627 (2012)
101. King, R., Brooks, S.P.: On the Bayesian analysis of population size. Biometrika **88**, 317–336 (2001)
102. Le, T., Clarke, B.: A Bayes interpretation of stacking for M-complete and M-open settings. Bayesian Anal. **12**, 807–829 (2017)
103. Lee, H.K.H.: Model selection for neural network classification. J. Classif. **18**, 227–243 (2001)
104. Ley, E., Steel, M.F.J.: Jointness in Bayesian variable selection with applications to growth regression. J. Macroecon. **29**, 476–493 (2007)
105. Ley, E., Steel, M.F.J.: On the effect of prior assumptions in Bayesian model averaging with applications to growth regression. J. Appl. Economet. **24**, 651–674 (2009)
106. Ley, E., Steel, M.F.J.: Comments on Jointness of growth determinants. J. Appl. Economet. **24**, 248–251 (2009)
107. Ley, E., Steel, M.F.J.: Mixtures of g-priors for Bayesian model averaging with economic applications. J. Economet. **171**, 251–266 (2012)
108. Li, G., Shi, J., Zhou, J.: Bayesian adaptive combination of short-term wind speed forecasts from neural network models. Renew. Energ. **36**, 352–359 (2011)
109. Li, Y., Clyde, M.A.: Mixtures of g-priors in generalized linear models. J. Am. Stat. Assoc. (2018). https://doi.org/10.1080/01621459.2018.1469992
110. Liang, F., Wong, W.H.: Evolutionary Monte Carlo: applications to C_p model sampling and change point problem. Stat. Sin. 317–342 (2000)
111. Liang, F., Paulo, R., Molina, G., Clyde, M.A., Berger, J.O.: Mixtures of g priors for Bayesian variable selection. J. Am. Stat. Assoc. **103**, 410–423 (2008)
112. Liddle, A.R.: Information criteria for astrophysical model selection. Mon. Not. R. Astron. Soc. **377**, L74–L78 (2007)
113. Lindley, D.V.: A statistical paradox. Biometrika **44**, 187–192 (1957)
114. Link, W., Barker, R.: Model weights and the foundations of multimodel inference. Ecology **87**, 2626–2635 (2006)
115. Link, W.A., Barker, R.J.: Bayesian Inference: With Ecological Applications. Academic Press, New York (2010)
116. Lu, D., Ye, M., Neuman, S.P.: Dependence of Bayesian model selection criteria and Fisher information matrix on sample size. Math. Geosci. **43**, 971–993 (2011)
117. Lumley, T., Scott, A.: AIC and BIC for modeling with complex survey data. J. Surv. Stat. Methodol. **3**, 1–18 (2015)
118. Madigan, D., Raftery, A.E.: Model selection and accounting for model uncertainty in graphical models using Occam's window. J. Am. Stat. Assoc. **89**, 1535–1546 (1994)
119. Madigan, D., York, J., Allard, D.: Bayesian graphical models for discrete data. Int. Stat. Rev. **63**, 215–232 (1995)
120. Maruyama, Y., George, E.I.: Fully Bayes factors with a generalized g-prior. Ann. Stat. **39**, 2740–2765 (2011)
121. Meng, X.-L., Wong, W.H.: Simulating ratios of normalizing constants via a simple identity: a theoretical exploration. Stat. Sin. **6**, 831–860 (1996)
122. Millar, R.B.: Comparison of hierarchical Bayesian models for overdispersed count data using DIC and Bayes' factors. Biometrics **65**, 962–969 (2009)
123. Min, C.-K., Zellner, A.: Bayesian and non-Bayesian methods for combining models and forecasts with applications to forecasting international growth rates. J. Econ. **56**, 89–118 (1993)
124. Minka, T.: Bayesian model averaging is not model combination. MIT Media Lab Note, December 2000
125. Mohammadi, A., Wit, E.C.: Bayesian structure learning in sparse Gaussian graphical models. Bayesian Anal. **10**, 109–138 (2015)

126. Monteith, K., Carroll, J.L., Seppi, K., Martinez, T.: Turning BMA into Bayesian model combination. In: International Joint Conference on Neural Networks (2011)

127. Moore, J.E., Barlow, J.: Bayesian statespace model of fin whale abundance trends from a 1991–2008 time series of linetransect surveys in the California Current. J. Appl. Ecol. **48**, 1195–1205 (2011)

128. Moral-Benito, E.: Determinants of economic growth: a Bayesian panel data approach. Rev. Econ. Stat. **94**, 566–579 (2012)

129. Moral-Benito, E.: Model averaging in economics: an overview. J. Econ. Surv. **29**, 46–75 (2015)

130. Müller, S., Scealy, J.L., Welsh, A.H.: Model selection in linear mixed models. Stat. Sci. **28**, 135–167 (2013)

131. Nelder, J.A., Wedderburn, R.W.M.: Generalized linear models. J. Roy. Stat. Soc. A. Stat. **135**, 370–384 (1972)

132. Newton, M.A., Raftery, A.E.: Approximate Bayesian inference with the weighted likelihood bootstrap. J. Roy. Stat. Soc. B. Methodol. **56**, 3–48 (1994)

133. Nott, D.J., Kohn, R.: Adaptive sampling for Bayesian variable selection. Biometrika **92**, 747–763 (2005)

134. O'Hagan, A.: Discussion of Aitkin, M.: Posterior Bayes factors. J. Roy. Stat. Soc. B. Methodol. **53**, 136 (1991)

135. O'Hagan, A.: Fractional Bayes factors for model comparison. J. Roy. Stat. Soc. B. Methodol. **57**, 99–138 (1995)

136. O'Hara, R.B., Mikko, J.S.: A review of Bayesian variable selection methods: what, how and which. Bayesian Anal. **4**, 85–117 (2009)

137. Parry, M.: Extensive scoring rules. Electron. J. Stat. **10**, 1098–1108 (2016)

138. Pauler, D.K.: The Schwarz criterion and related methods for normal linear models. Biometrika **85**, 13–27 (1998)

139. Pauler, D.K., Wakefield, J.C., Kass, R.E.: Bayes factors and approximations for variance component models. J. Am. Stat. Assoc. **94**, 1242–1253 (1999)

140. Pérez, J.M., Berger, J.O.: Expected-posterior prior distributions for model selection. Biometrika **89**, 491–511 (2002)

141. Pole, A., West, M., Harrison, J.: Applied Bayesian Forecasting and Time Series Analysis. CRC Press, Boca Raton (1994)

142. Pooley, C.M., Marion, G.: Bayesian model evidence as a practical alternative to deviance information criterion. Roy. Soc. Open Sci. **5**, 171519 (2018)

143. Price, M.J., Welton, N.J., Briggs, A.H., Ades, A.E.: Model averaging in the presence of structural uncertainty about treatment effects: influence on treatment decision and expected value of information. Value Health **14**, 205–218 (2011)

144. R Core Team: R: A language and environment for statistical computing. R Foundation for Statistical Computing, Vienna, Austria (2017). https://www.R-project.org/

145. Raftery, A.E.: Bayesian model selection in social research. Sociol. Methodol. **25**, 111–164 (1995)

146. Raftery, A.E.: Approximate Bayes factors and accounting for model uncertainty in generalised linear models. Biometrika **83**, 251–266 (1996)

147. Raftery, A.E., Madigan, D., Hoeting, J.A.: Bayesian model averaging for linear regression models. J. Am. Stat. Assoc. **92**, 179–191 (1997)

148. Raftery, A.E., Zheng, Y.: Discussion of Hjort, N.L., Claeskens, G.: Frequentist model average estimators. J. Am. Stat. Assoc. **98**, 931–938 (2003)

149. Raftery, A.E., Kárný, M., Ettler, P.: Online prediction under model uncertainty via dynamic model averaging: application to a cold rolling mill. Technometrics **52**, 52–66 (2010)

150. Rissanen, J.: A universal prior for integers and estimation by minimum description length. Ann. Stat. **11**, 416–431 (1983)

151. Robert, C.P., Marin, J.-M.: On some difficulties with a posterior probability approximation technique. Bayesian Anal. **3**, 427–441 (2008)

152. Rossell, D., Telesca, D.: Nonlocal priors for high-dimensional estimation. J. Am. Stat. Assoc. **112**, 254–265 (2017)
153. Sabanés Bové, D., Held, L.: Bayesian fractional polynomials. Stat. Comput. **21**, 309–324 (2011)
154. Sabanés Bové, D., Held, L.: Hyper-g priors for generalized linear models. Bayesian Anal. **6**, 387–410 (2011)
155. Sabanés Bové, D., Held, L., Kauermann, G.: Objective Bayesian model selection in generalized additive models with penalized splines. J. Comput. Graph. Stat. **24**, 394–415 (2015)
156. Sala-i-Martin, X., Doppelhofer, G., Miller, R.: Determinants of long-term growth: a Bayesian averaging of classical estimates (BACE) approach. Am. Econ. Rev. **94**, 813–835 (2004)
157. Schwarz, G.: Estimating the dimension of a model. Ann. Stat. **6**, 461–464 (1978)
158. Scott, J.G., Berger, J.O.: Bayes and empirical-Bayes multiplicity adjustment in the variable-selection problem. Ann. Stat. **38**, 2587–2619 (2010)
159. Spiegelhalter, D.J., Best, N.G., Carlin, B.P., van der Linde, A.: Bayesian measures of model complexity and fit. J. R. Stat. Soc. B. Methodol. **64**, 583–639 (2002)
160. Spiegelhalter, D.J., Best, N.G., Carlin, B.P., van der Linde, A.: The deviance information criterion: 12 years on. J. R. Stat. Soc. B. Methodol. **76**, 485–493 (2014)
161. Steel, M.F.J.: Bayesian model averaging and forecasting. Bull. EU US Inflation Macroecon. Anal. **200**, 30–41 (2011)
162. Steel, M.F.J.: Model averaging and its use in economics (2017). arXiv preprint: arXiv:1709.08221
163. Stock, J.H., Watson, M.W.: Forecasting with many predictors. In: Elliott, C.G.G., Timmermann, A. (eds.) Handbook of Economic Forecasting. Elsevier (2006)
164. Stone, M.: Cross-validatory choice and assessment of statistical predictions. J. Roy. Stat. Soc. B. Methodol. **36**, 111–147 (1974)
165. Stone, M.: Comments on model selection criteria of Akaike and Schwarz. J. Roy. Stat. Soc. B. Methodol. **41**, 276–278 (1979)
166. Strachan, R.W.: Comment on Jointness of growth determinants by Gernot Doppelhofer and Melvyn Weeks. J. Appl. Economet. **24**, 245–247 (2009)
167. Thogmartin, W.E., Knutson, M.G., Sauer, J.R.: Predicting regional abundance of rare grassland birds with a hierarchical spatial count model. Condor **108**, 25–46 (2006)
168. Vehtari, A., Gelman, A., Gabry, J.: Pareto smoothed importance sampling (2017). arXiv preprint: arxiv:1507.02646
169. Vehtari, A., Gelman, A., Gabry, J.: Practical Bayesian model evaluation using leave-one-out cross-validation and WAIC. Stat. Comput. **27**, 1413–1432 (2017)
170. Villa, C., Walker, S.: An objective Bayesian criterion to determine model prior probabilities. Scand. J. Stat. **42**, 947–966 (2015)
171. Volinsky, C.T., Raftery, A.E.: Bayesian information criterion for censored survival models. Biometrics **56**, 256–262 (2000)
172. Walker, S.G., Gutiérrez-Peña, E., Muliere, P.: A decision theoretic approach to model averaging. J. Roy. Stat. Soc. D Stat. **50**, 31–39 (2001)
173. Wang, C., Dominici, F., Parmigiani, G., Zigler, C.M.: Accounting for uncertainty in confounder and effect modifier selection when estimating average causal effects in generalized linear models. Biometrics **71**, 654–665 (2015)
174. Watanabe, S.: Asymptotic equivalence of Bayes cross validation and widely applicable information criterion in singular learning theory. J. Mach. Learn. Res. **11**, 3571–3594 (2010)
175. Watanabe, S.: A widely applicable Bayesian information criterion. J. Mach. Learn. Res. **14**, 867–897 (2013)
176. Wei, Y., McNicholas, P.D.: Mixture model averaging for clustering. Adv. Data Anal. Classi. **9**, 197–217 (2015)
177. Wilberg, M.J., Bence, J.R.: Performance of deviance information criterion model selection in statistical catch-at-age analysis. Fish. Res. **93**, 212–221 (2008)
178. Wong, H., Clarke, B.: Improvement over Bayes prediction in small samples in the presence of model uncertainty. Can. J. Stat. **32**, 269–283 (2004)

179. Xie, W., Lewis, P.O., Fan, Y., Kuo, L., Chen, M.-H.: Improving marginal likelihood estimation for Bayesian phylogenetic model selection. Syst. Biol. **60**, 150–160 (2010)
180. Yang, Y.: Can the strengths of AIC and BIC be shared? A conflict between model indentification and regression estimation. Biometrika **92**, 937–950 (2005)
181. Yao, Y., Vehtari, A., Simpson, D., Gelman, A.: Using stacking to average Bayesian predictive distributions. Bayesian Anal. (2018). https://doi.org/10.1214/17-BA1091
182. Yao, Y., Vehtari, A., Simpson, D., Gelman, A.: Rejoinder to the Discussion of Yao, Y., Vehtari, A., Simpson, D., Gelman, A.: Using stacking to average Bayesian predictive distributions. Bayesian Anal. (2018). https://doi.org/10.1214/17-BA1091
183. Ye, M., Meyer, P.D., Neuman, S.P.: On model selection criteria in multimodel analysis. Water Resour. Res. **44**, W03428 (2008)
184. Yeung, K.Y., Bumgarner, R.E., Raftery, A.E.: Bayesian model averaging: development of an improved multi-class, gene selection and classification tool for microarray data. Bioinformatics **21**, 2394–2402 (2005)
185. Zaffalon, M.: The naive credal classifier. J. Stat. Plan. Infer. **105**, 5–21 (2002)
186. Zellner, A., Siow, A.: Posterior odds ratios for selected regression hypotheses. In: Bernardo, J.M., DeGroot, M.H., Lindley, D.V., Smith, A.F.M. (eds.) Bayesian Statistics: Proceedings of the First International Meeting held in Valencia, Spain, pp. 585–603. University Press (1980)
187. Zellner, A.: On assessing prior distributions and Bayesian regression analysis with g-prior distributions. In: Goel, P.K., Zellner, A. (eds.) Bayesian Inference and Decision Techniques: Essays in Honor of Bruno De Finetti, pp. 233–243. Elsevier Science, Oxford (1986)
188. Zhao, J., Jin, L., Shi, L.: Mixture model selection via hierarchical BIC. Comput. Stat. Data Anal. **88**, 139–153 (2015)
189. Zigler, C.M., Dominici, F.: Uncertainty in propensity score estimation: Bayesian methods for variable selection and model-averaged causal effects. J. Am. Stat. Assoc. **109**, 95–107 (2014)

Chapter 3
Frequentist Model Averaging

Abstract We provide an overview of frequentist model averaging. For point estimation, we consider different methods for selecting the model weights, including those based on AIC, bagging, weighted AIC, stacking and focussed methods. For interval estimation, we consider Wald, MATA and percentile-bootstrap intervals. Use of the methods are illustrated by examples involving real data.

3.1 Introduction

We now consider how frequentist model averaging (FMA) can be used to obtain point estimates and confidence intervals for θ. As we shall see, there are clear links between some of the methods and BMA.

3.2 Point Estimation

Suppose we have a set of model weights $w = \{w_1, \ldots, w_M\}$ that lie on the unit simplex. A model-averaged point estimate of θ is given by

$$\widehat{\theta} = \sum_{m=1}^{M} w_m \widehat{\theta}_m, \tag{3.1}$$

where $\widehat{\theta}_m$ is the estimate obtained from model m. The estimate in (3.1) is the frequentist analogue of the model-averaged posterior mean in (2.4).

In many situations there will be a natural scale on which to use the arithmetic weighted mean in (3.1), namely that on which the effects of the predictor variables are assumed to be additive. When averaging over a set of GLMs, for example, it is natural to do so on the linear predictor scale (Sects. 1.3.1 and 2.3.1). An interesting example arises when we have a set of Poisson models, such as in the sea lion example introduced in Sect. 1.3.1. If we want to estimate μ, the expected value of Y for specific

© The Author(s), under exclusive licence to Springer-Verlag GmbH, DE, part of Springer Nature 2018

D. Fletcher, *Model Averaging*, SpringerBriefs in Statistics, https://doi.org/10.1007/978-3-662-58541-2_3

values of the predictor variables, it is natural to use (3.1) with $\theta = \log \mu$, and then obtain a model-averaged estimate of μ as

$$\widehat{\mu} = e^{\widehat{\theta}} = \prod_{m=1}^{M} \widehat{\mu}_m^{w_m},$$

where $\widehat{\mu}_m = e^{\widehat{\theta}_m}$ is the estimate of μ obtained from model m. Thus $\widehat{\mu}$ is a geometric weighted mean on the original scale. In what follows, θ is taken to be the parameter of interest after transformation to the scale on which averaging is most natural.

This example illustrates the fact that a frequentist model-averaged point estimate is not transformation-invariant, except in the uninteresting case where the transformation is linear. Likewise, in BMA, the model-averaged posterior distribution does not depend on the parametrisation, but the model-averaged posterior mean does (Sect. 2.3.1).

As the model weights are estimated from the data, and the estimates of θ from the different models may be correlated, the sampling distribution of $\widehat{\theta}$ will be difficult to assess analytically [21, 38, 88, 121]. This makes derivation of a reliable model-averaged confidence interval difficult (Sect. 3.4).

We now consider a range of approaches to selecting the model weights in FMA. It is important to note that none of these methods is guaranteed to be optimal, in terms of the frequentist properties of $\widehat{\theta}$ [38, 113]. In addition, estimation of the weights may not be reliable when n is small (Sect. 1.4).

3.2.1 Information-Criterion Methods

A common choice of model weight in FMA is the AIC weight, given by

$$w_m \propto \exp\left(-\text{AIC}_m/2\right), \tag{3.2}$$

where AIC_m is Akaike's information criterion for model m [2, 3, 159], given by

$$\text{AIC}_m = -2 \log L\left(\widehat{\beta}_m \middle| y, m\right) + 2p_m, \tag{3.3}$$

where $L\left(\beta_m \mid y, m\right)$ is the likelihood under model m and $\widehat{\beta}_m$ is the maximum likelihood estimate for β_m. Note the change in notation from the Bayesian setting, where $p\left(y \mid \beta_m, m\right)$ was used to denote the likelihood in (2.3). Use of (3.2) is motivated by analogy with the BIC weight in (2.10) [4, 5, 21].

AIC is an approximately unbiased estimate of the expected relative Kullback-Leibler divergence between the data-generating mechanism and the fitted model. Like BIC, it can be thought of as providing a trade-off between model-fit and model-complexity, corresponding to the first and second terms on the right-hand side of (3.3).

Interestingly, derivation of the $2p_m$ correction for overfitting involves the assumption that model m is true [38, 159].[1]

AIC is a prediction-based criterion, whereas BIC is more concerned with identification of the true model [159] (Sect. 1.4). We might therefore expect AIC weights to perform better than those based on BIC, particularly if θ is the expected value of Y for specific values of the predictor variables.

Comparison of (2.9) and (3.2) shows that the AIC weight is a special case of the generalised BIC weight, corresponding to a prior probability for model m of the form

$$p\left(m\right) \propto \exp\left\{p_m \left(\frac{1}{2}\log n - 1\right)\right\}. \tag{3.4}$$

This prior will give more weight to the larger models as n increases, a well-known feature of AIC [24, 38, 124]. This result has connections with the work of [206], who suggested that classical BMA will not have good frequentist prediction properties unless we allow the model-prior to depend on n (Sect. 2.5). In related work, [182] reported the results of a simulation study which showed that use of the prior in (3.4) led to better frequentist coverage rates than a uniform model-prior.

For a mixed effects model, it is not always clear what form AIC should take, as the value of p_m is not clear-cut, and the focus of the analysis needs to be defined. We discuss the mixed-model setting further in Sect. 3.6.

A small-sample modification to AIC, proposed by [177] in the context of normal linear models[2] and autoregressive models, is given by

$$\text{AICc}_m = -2\log L\left(\widehat{\beta}_m \big| y, m\right) + 2p_m\left(\frac{n}{n - p_m - 1}\right). \tag{3.5}$$

Replacing AIC_m by AICc_m in (3.2) provides the AICc weight for model m. As with BIC weights, some care is needed when using AICc weights if the definition of n is not obvious [24] (Sect. 2.2.1). A useful overview of the differences between AIC and AICc is provided by [29].

The corrections for overfitting used by AIC, AICc and BIC in (3.3), (3.5) and (2.8), indicate that BIC will give less weight to larger models than AIC when $n \geq 8$, and that AICc will always do so. For small n, AICc gives less weight to larger models than BIC, while for large n AICc and AIC are almost identical.

Interestingly, when we have two nested models, and $p_2 = p_1 + 1$, the maximum possible AIC weight for the smaller model is $e/(1 + e) \approx 0.73$ [22, 187]. This can be viewed as an inherent drawback to the use of AIC weights, or as a sign that there

---

[1] This can come as a surprise; see [159] for a useful discussion of the assumptions underlying AIC.

[2] As discussed in Sect. 2.2, when counting the number of parameters in a model we include any scale parameters, such as the error variance in a normal linear model.

is a natural limit to the weight that should be given to the smaller model, for the purpose of estimation (Sect. 1.4).

Use of AICc weights was proposed by [24], who suggested that they should work well if the likelihood function is close to what it would be for a normal linear model. As suggested by [38], AICc should be used with care outside the settings for which it was developed (normal linear models and autoregressive models) [158]. Much of the discussion in the literature on AICc is concerned with model selection rather than model averaging [23, 95, 96, 140]. For model-averaged interval estimation, there is evidence to suggest that AIC weights are preferable to those based on AICc, even in the normal linear model setting [27, 59, 107]. In the examples, we therefore consider weights based on AIC rather than AICc.

In the context of GLMs, the following adjustment to AIC has been suggested when using quasi-likelihood to allow for overdispersion [87, 196]:

$$\text{QAIC}_m = \frac{-2 \log L \left(\widehat{\beta}_m \middle| y, m \right)}{\widehat{\phi}_M} + 2 p_m, \tag{3.6}$$

where $\widehat{\phi}_M$ is the estimate of overdispersion obtained from model M [9, 61]. An alternative version of QAIC is given by

$$\text{QAIC}_m = -2 \log L \left(\widehat{\beta}_m \middle| y, m \right) + 2 p_m \widehat{\phi}_m, \tag{3.7}$$

where $\widehat{\phi}_m$ is the estimate of overdispersion obtained from model m [38]. A modification of AIC similar to that in (3.7) was proposed by [133], in the related setting of allowing for a design effect when analysing survey data (see also [92]). Use of either version of QAIC in (3.2) will lead to weights that are lower for larger models than the corresponding AIC weights.

In principle, we could replace AIC in (3.2) by one of several alternative criteria, such as the risk inflation criterion (RIC) or the Kullback information criterion (KIC) [31, 40, 62, 72]. Some criteria involve data-dependent corrections for overfitting, such as Takeuchi's Information Criterion (TIC), which can be thought of as a frequentist version of DIC. Unlike AIC it does not require the assumption that the model is true [38, 159, 178], but calculation of the correction for overfitting is not straightforward and may be prone to instability if n is small [24, 38]. Other data-dependent corrections for overfitting have been proposed, with a view to providing a compromise between AIC, BIC and RIC [14, 72, 78]. The network information criterion (NIC), proposed by [145] for selection of neural networks, is a generalisation of TIC to situations in which model-fit is not based on maximum likelihood.

One disadvantage of weights based on an information criterion is that model-redundancy can lead to some of the weights being inappropriately diluted [24], similar to the issues that can arise for the model-prior in classical BMA (Sect. 2.2.2).

3.2.2 Bagging

Bagging, also known as bootstrap-aggregating or bootstrap-smoothing, involves using the bootstrap to mimic the process of model selection [16, 55, 84]. Thus we generate B bootstrap samples and for each of these we note the estimate of θ obtained from the best model for that sample. The choice of best model can be based on any criterion, such as AIC. The original motivation for bagging came from the potential for instability in model selection when using classification trees in machine learning, as a small change in the training data can lead to a large change in the choice of best classifier [16, 17].

If we assume that model M is true, it is natural to generate the bootstrap samples from this model, as in other bootstrap-based approaches to model averaging [21, 55, 130], but one could use a non-parametric bootstrap [16, 21, 55, 84, 130]. Regardless of the method for generating bootstrap samples, the model-averaged bagging estimate is given by

$$\frac{1}{B}\sum_{b=1}^{B}\widehat{\theta}_{(b)}, \tag{3.8}$$

where $\widehat{\theta}_{(b)}$ is the estimate from the best model for bootstrap sample b. The expression in (3.8) can also be written as

$$\sum_{m=1}^{M} w_m \bar{\theta}_m^{BAG}, \tag{3.9}$$

where $\bar{\theta}_m^{BAG}$ is the mean of $\widehat{\theta}_m$ from the B_m samples in which model m is selected as the best and $w_m = B_m/B$ is the proportion of times that model m is selected as the best. Thus (3.8) can be regarded as an estimate of

$$\sum_{m=1}^{M} p\,(S = m)\,\mathrm{E}\left(\widehat{\theta}_m \Big| S = m\right), \tag{3.10}$$

where S is a random variable denoting the model selected as the best.[3] Comparison of (3.10) with (2.4) shows the connection between bagging and the posterior mean in BMA [84, 157], with the concept that a model is true being replaced by the concept that it is selected as the best. Note that even if we use AIC to select the best model for each bootstrap sample, the estimate in (3.8) will differ from that obtained using AIC weights; the bycatch example in Sect. 3.3.1 illustrates this point.

One advantage of bagging over an information-criterion-based method arises when we have some model-redundancy (Sect. 2.2.2). In the extreme case where two models have almost identical likelihoods, regardless of the data, bagging will select just one of these as the best for each bootstrap sample. The weights for these two

---

[3] See [24] for a discussion of the connection between the model-selection probabilities $p\,(S = m)$ $(m = 1, \ldots, M)$ and AIC weights.

models in (3.9) will therefore be such that their total is approximately the same as the weight we would obtain for just one of them if we recognised the redundancy and omitted the other model from consideration [21]. This is clearly not true for a model weight based on an information criterion.

3.2.3 Optimal Weights

As discussed in Sect. 1.2, use of a model-averaged point estimate can be regarded as a means of obtaining a good balance between bias and variance, as in model selection. This perspective has been the focus of recent research into selection of an optimal set of weights, based on an objective function. The estimation of these optimal weights involves some form of constrained-optimisation, exactly as in BSP (Sect. 2.3.2).[4] We consider two approaches concerned with prediction of a new value of y: a weighted version of AIC and frequentist stacking. For both of these, we focus on the GLM setting, but the ideas should apply in many other settings. We also consider a focussed approach, which is concerned directly with the sampling properties of $\widehat{\theta}$.

AIC(w)
Suppose we wish to average over a set of GLMs. As discussed in Sects. 1.3.1, 2.3.1 and 3.2, it is natural to perform such averaging on the linear predictor scale, and we assume that is the case here. If we assess models purely on a measure of the within-sample prediction error, such as the deviance, we will give all the weight to model M [84, 111]. The prediction error will also be underestimated, as the data are being used to both fit the models and to estimate their prediction error, exactly the same overfitting problem we encounter when using a single model. To adjust for this underestimation, we can add a correction for overfitting to the objective function, as in AIC.

Let η_i be the linear predictor for observation i and

$$\widehat{\eta}_i = \sum_{m=1}^{M} w_m \widehat{\eta}_{mi} \tag{3.11}$$

be the model-averaged estimate of η_i, where $\widehat{\eta}_{mi}$ is the maximum likelihood estimate of η_i obtained from fitting model m. A model-weighted version of AIC in this setting was proposed by [223] and is given by

$$\text{AIC}(w) = -2 \sum_{i=1}^{n} \log L\left(\widehat{\eta}_i \Big| y_i\right) + 2p(w), \tag{3.12}$$

---

[4]Throughout the rest of the chapter, it will be implicit that constrained-optimisation is used whenever we determine the weights by minimising an objective function.

where $\log L(\eta_i \mid y_i)$ is the contribution from y_i to the log-likelihood, which is now written as a function of η_i. The correction for overfitting involves the weighted average of the number of parameters in each model, i.e.

$$p(w) = \sum_{m=1}^{M} w_m p_m.$$

Note that the likelihood in (3.12) is not the maximised likelihood under a particular model, unlike in AIC. The optimal weights are taken to be those that minimise $\text{AIC}(w)$, in the same way that minimisation of AIC is used in model selection.

When we have a scale parameter, as in the normal linear model setting, this needs to be included in the likelihood term in (3.12). Typically, this parameter is estimated using the largest model [79], but the choice of estimate is not important for the selection of weights. Likewise, we could define p_m in (3.12) to be the number of regression coefficients, rather than the total number of parameters, as the choice of weights will not be affected [223].

This approach can be thought of as equivalent to fitting a generalised linear meta-model to y, with the predictor variables being $\widehat{\eta}_{1i}, \ldots, \widehat{\eta}_{Mi}$, the regression coefficients being the model weights, and the model being fitted by minimising $\text{AIC}(w)$. As with standard regression models, it will be good practice to check for collinearity in the $\widehat{\eta}_{1i}, \ldots, \widehat{\eta}_{Mi}$ before minimising (3.12). In particular, if two models give very similar predictions, we have some model-redundancy, and it may be sensible to exclude one of them before proceeding [24]. Even if this problem is overlooked, the constraint that the model weights be non-negative will usually lead to one of the models being given very little weight.

Use of the alternative correction term $p(w) \log n$ in (3.12) gives $\text{BIC}(w)$, a model-weighted version BIC [223]. If there is overdispersion, a quasi-likelihood version of $\text{AIC}(w)$ could be used, based on either (3.6) or (3.7). Simulation results in [223] showed that $\text{AIC}(w)$ and $\text{BIC}(w)$ can both outperform AIC and BIC weights (Sect. 3.2.1). Similar results, using both asymptotic theory and simulation, were provided by [79] for $\text{AIC}(w)$ when averaging over a set of linear models.[5]

Stacking

Stacking is the model-averaging equivalent of cross validation for model selection [175]. As it involves estimating the out-of-sample prediction error, there is no correction for overfitting. For averaging a set of GLMs [11], the stacking weights[6] are those that maximise

$$\sum_{i=1}^{n} \log L\left(\widehat{\eta}_i \mid y_i\right), \tag{3.13}$$

---

[5]For normal linear models, $\text{AIC}(w)$ is equivalent to Mallows model averaging (MMA) [79, 121, 143, 190, 220, 223]. Although MMA was developed without the assumption of normal errors, for simplicity we use the more general name $\text{AIC}(w)$ when referring to MMA.

[6]As with $\text{AIC}(w)$, the choice of estimate of any scale parameter will not affect the weights.

where we now have

$$\widehat{\eta}_i = \sum_{m=1}^{M} w_m \widehat{\eta}_{m[-i]}, \tag{3.14}$$

and $\widehat{\eta}_{m[-i]}$ is the estimate of η_i obtained from fitting model m to all the data except y_i. The special case of high-dimensional linear models was considered by [10]. As in the Bayesian setting (Sect. 2.3.2), frequentist stacking does not involve assuming a particular form for the true data-generating mechanism.

As pointed out by [111], stacking was originally developed by [175] (who referred to it as a model-mix prescription) in the context of averaging over a set of normal linear models. It was re-discovered in machine learning, where the term stacking originated [15, 84, 99, 111, 180, 198] and later in econometrics, where it has been referred to as jackknife model averaging [83].[7]

As with AIC(w), selecting weights by maximisation of (3.13) is equivalent to fitting a generalised linear meta-model to y, this time with predictor variables $\widehat{\eta}_{1[-i]}, \ldots, \widehat{\eta}_{M[-i]}$ [45]. The same remarks apply regarding possible collinearity in these predictors, and the benefit of having a non-negativity constraint on the weights [45].

When n is large, stacking might be expected to produce weights that are similar to those based on AIC and AIC(w), by analogy with the result that AIC and leave-one-out cross validation are asymptotically equivalent [176]. For linear models with non-constant error variance, stacking has been shown to perform better than weights based on AIC(w), AIC or BIC [83]. However, a simple modification to the correction for overfitting in AIC(w) can be used to allow for a non-constant error variance [125], and simulation results suggest that this leads to the same performance as stacking.[8] A number of asymptotic results for stacking have been provided [83, 219], analogous to those for cross validation [12, 140].

As in model selection, we could use k-fold cross validation ($k < n$), in which n/k observations are omitted from the data set each time [159]. A number of authors have discussed the issues involved in choosing k [84, 159, 222]; for simplicity we focus on the leave-one-out approach ($k = n$). A number of other sample-splitting techniques have been proposed in the context of model averaging, but we do not consider these in detail [21, 123, 126, 204, 205, 212, 213].

Focussed methods

If θ is the expected value of Y for specific values of the predictor variables, we might expect a prediction-based method such as AIC(w) or stacking to provide sensible weights. In general, however, we might prefer a method that is explicitly tailored to the choice of θ. Focussed methods involve determining the weights in order to minimise an estimate of the error associated with $\widehat{\theta}$, with different choices of θ typically leading to different model weights.

---

[7]This name is potentially confusing as the original jackknife is somewhat different, involving the use of pseudo-values to reduce the bias of an estimate obtained from a single model [48, 154, 175].

[8]Other modifications to AIC(w) in this setting have been proposed [129, 220, 226].

This approach was considered by [88], who suggested finding the model weights that minimise an estimate of the MSE of $\widehat{\theta}$. In some settings a measure of estimation error other than MSE might be appropriate, such as classification error rate [36]. In deriving the estimate of the MSE, [88] assumed that the model weights are constant; this is clearly not true, but simplifies the theory.[9] They also assumed that the models are nested, the largest model is true, and the difference between the largest and smallest model vanishes with n. The last of these assumptions is referred to as local misspecification, and there has been some discussion of the need for it in the literature [88, 89, 103, 156].

Recently, [93] has used the results of [197] on maximum likelihood estimation under model-misspecification to develop a focussed method for GLMs that does not rely on the local-misspecification assumption. Simulation results show that this alternative can perform better than the local-misspecification method of [88]. A similar approach, that also does not require local-misspecification, has recently been proposed by [142], in a more general likelihood-model setting.

If we wish to avoid using asymptotic theory, we can calculate a parametric bootstrap-based estimate of the MSE of $\widehat{\theta}$ as follows. Assuming the model weights are constant, this MSE is given by

$$\text{MSE}\left(\widehat{\theta}\right) = \sum_{m_1=1}^{M} \sum_{m_2=1}^{M} w_{m_1} w_{m_2} e_{m_1 m_2}, \tag{3.15}$$

where

$$e_{m_1 m_2} = \text{E}\left\{\left(\widehat{\theta}_{m_1} - \theta\right)\left(\widehat{\theta}_{m_2} - \theta\right)\right\},$$

and expectation is with respect to model M. The model weights can be obtained by minimising a bootstrap-based estimate of (3.15) as follows. For $j = 1, \ldots, B$, we randomly generate data $y^{(j)}$ from the fitted version of model M, and calculate $\widehat{\theta}_m^{(j)}$ by fitting model m to $y^{(j)}$. We then estimate $e_{m_1 m_2}$ by

$$\widehat{e}_{m_1 m_2} = \frac{1}{B} \sum_{j=1}^{B} \left(\widehat{\theta}_{m_1}^{(j)} - \widehat{\theta}_M\right)\left(\widehat{\theta}_{m_2}^{(j)} - \widehat{\theta}_M\right).$$

An added advantage of using a bootstrap-based approach is that we can avoid assuming that $\widehat{\theta}_M$ is unbiased, unlike methods which use asymptotic approximations. In the examples below, we therefore focus on use of this bootstrap-based approach.

The actual MSE of $\widehat{\theta}$ will be higher than that predicted by (3.15), as the model weights are clearly not fixed. This is not a problem if the method provides a good indication of the relative size of the true MSE for different choices of the weights.

---

[9]This assumption has also been used in interval estimation (Sect. 3.4.1).

3.3 Examples

3.3.1 Sea Lion Bycatch

For the sea lion bycatch example, the AIC values for models 1 and 2 are 12.95 and 15.41 respectively. Use of (3.2) then leads to AIC weights of 0.773 and 0.227 respectively, as we saw in Sect. 1.3.1.

Suppose we now carry out bagging, using 10^4 bootstrap samples generated from model 2, and for each bootstrap sample we select the best model using AIC. The two types of model-averaged estimate are shown in Table 3.1, together with those for each model. Both of the model-averaged estimates provide a compromise between the estimates from the two models. As discussed in Sect. 3.2.2, the bagging estimate is a weighted mean of $\bar{\theta}_1^{BAG}$ and $\bar{\theta}_2^{BAG}$, where $\bar{\theta}_m^{BAG}$ is the mean of the estimates from the bootstrap samples in which model m is the best. In this case, to the nearest integer, we have $\bar{\theta}_1^{BAG} = 34, 117$ and 4 sea lions for scampi, squid and other species respectively; the corresponding values for $\bar{\theta}_2^{BAG}$ are 78, 95 and 17 sea lions. The weights implicit in the use of bagging are $B_1/B = 0.642$ and $B_2/B = 0.358$.

Compared to the use of AIC weights, bagging gives a higher estimate of total bycatch for vessels targeting scampi and other species, and a lower estimate for those targeting squid. These differences arise because the two types of model weight differ and, for all three fisheries, the mean of the estimates obtained when model 2 was selected as the best differs substantially from that obtained by fitting model 2 to the original data.

3.3.2 Ecklonia Density

The AIC values for models 1 and 2 are 862.4 and 860.5 respectively, and use of (3.2) leads to AIC weights of 0.276 and 0.724 respectively, as we saw in Sect. 1.3.2.

If we use AIC(w), $\widehat{\eta}_{1i}$ and $\widehat{\eta}_{2i}$ are uncorrelated, as $\widehat{\eta}_{1i}$ is the same for all i ($i = 1, \ldots, n$). Likewise, for stacking $\widehat{\eta}_{1[-i]}$ and $\widehat{\eta}_{2[-i]}$ are uncorrelated. This lack of correlation will help avoid any numerical issues in optimising the functions in (3.12) and (3.13), analogous to estimation of the regression coefficients in a model with uncorrelated predictors.

Table 3.1 Estimates of total sea lion bycatch (to the nearest integer) obtained from each of two models and by model averaging using AIC weights and bagging

Species	Model 1	Model 2	Model-averaged	
			AIC	Bagging
Scampi	35	58	39	50
Squid	119	105	116	109
Other	4	10	5	9

Table 3.2 Model weights using AIC, AIC(w), stacking and bootstrap-based focussed model averaging for the ecklonia density study

Model	AIC	AIC(w)	Stacking	Focussed model averaging		
				Zone 1	Zone 2	Zone 3
1	0.276	0.306	0.201	0.180	0.969	0.168
2	0.724	0.694	0.799	0.820	0.031	0.832

For both AIC(w) and stacking, we used the estimate of k from the largest model when calculating the likelihood terms in (3.12) and (3.13). For bootstrap-based focussed model averaging (Sect. 3.2.3), we considered estimation of the true mean density in each zone, and used 10^4 bootstrap samples. The full set of weights is shown in Table 3.2.

The AIC, AIC(w) and stacking weights are similar and show a preference for model 2, as do the focussed weights for zones 1 and 3. For zone 2, however, focussed model averaging gives almost all the weight to model 1. To see why, consider the expectation term in (3.15). This can be written as

$$e_{m_1 m_2} = b_{m_1 m_2} + c_{m_1 m_2},$$

where

$$b_{m_1 m_2} = \text{bias}\left(\widehat{\theta}_{m_1}\right) \text{bias}\left(\widehat{\theta}_{m_2}\right)$$

and

$$c_{m_1 m_2} = \text{cov}\left(\widehat{\theta}_{m_1}, \widehat{\theta}_{m_2}\right).$$

The bootstrap-based estimates of these two terms are

$$\widehat{b}_{m_1 m_2} = \left(\bar{\theta}_{m_1} - \widehat{\theta}_M\right)\left(\bar{\theta}_{m_2} - \widehat{\theta}_M\right)$$

and

$$\widehat{c}_{m_1 m_2} = \frac{1}{B} \sum_{j=1}^{B} \left(\widehat{\theta}_{m_1}^{(j)} - \bar{\theta}_{m_1}\right)\left(\widehat{\theta}_{m_2}^{(j)} - \bar{\theta}_{m_2}\right),$$

where

$$\bar{\theta}_{m_i} = \frac{1}{B} \sum_{j=1}^{B} \widehat{\theta}_{m_i}^{(j)} \quad (i = 1, 2).$$

The estimate of the MSE in (3.15) can therefore be written as

$$\sum_{m_1=1}^{M} \sum_{m_2=1}^{M} w_{m_1} w_{m_2} \left\{ \widehat{b}_{m_1 m_2} + \widehat{c}_{m_1 m_2} \right\}. \tag{3.16}$$

For the bootstrap samples used to determine the focussed weights for zone 2 in Table 3.2, the estimates of the bias and covariance terms are

$$\widehat{b}_{11} = 0.0003, \quad \widehat{b}_{12} \equiv \widehat{b}_{21} = 0.0009, \quad \widehat{b}_{22} = 0.0027,$$

$$\widehat{c}_{11} = 0.0308, \quad \widehat{c}_{12} \equiv \widehat{c}_{21} = 0.0273, \quad \widehat{c}_{22} = 0.1143.$$

As we saw in Sect. 1.3.2, the estimates of density in zone 2 are almost identical for the two models, which leads to the estimates of the bias terms all being small relative to those for the covariance terms. This means that the choice of weights is largely determined by the estimates of the covariance terms. The largest term is $\widehat{c}_{22}$, reflecting the fact that model 2 contains more parameters. Minimisation of (3.16) therefore leads to giving most of the weight to model 1. For zones 1 and 3, the bias terms are more influential, the largest being $\widehat{b}_{11}$, and most of the weight is given to model 2.

The model-averaged estimates of mean density are shown in Table 3.3, together with the estimates from the two models, each estimate being obtained by back-transformation of that for $\log \mu_i$ ($i = 1, 2, 3$). For each zone, the model-averaged estimates are similar. For zones 1 and 3, this is because the three types of model weight are roughly the same. For zone 2, the two models give almost identical estimates, and the choice of weights therefore has little effect. The impact of choosing focussed weights is clearer when we consider interval estimation for this example in Sect. 3.5.2.

3.3.3 *Water-Uptake in Amphibia*

In Sect. 1.3.3, we considered use of AIC weights for the factorial experiment on water-uptake in amphibia. For comparison, we now consider the AICc and BIC weights as well (Table 3.4).

Table 3.3 Estimates of mean density of ecklonia (individuals per quadrat) in three zones for each of four methods of model averaging and for each model

Zone	Model 1	Model 2	Model-averaged			
			AIC	AIC(w)	Stacking	Focussed
1	33.5	16.9	20.4	20.8	19.4	19.1
2	33.5	33.6	33.6	33.6	33.6	33.5
3	33.5	45.3	41.7	41.3	42.6	43.0

Table 3.4 A set of candidate models for the water-uptake experiment, with model weights calculated using AIC, AICc and BIC. Weights larger than 0.1 are shown in bold

Model	AIC	AICc	BIC
Null	0.000	0.009	0.001
S	0.000	0.028	0.002
C	0.000	0.021	0.002
H	0.000	0.005	0.000
S+C	0.006	**0.185**	0.021
S+H	0.001	0.021	0.002
C+H	0.001	0.015	0.002
S+C+H	0.030	**0.267**	0.066
S+C+SC	0.003	0.031	0.008
S+H+SH	0.001	0.009	0.002
C+H+CH	0.000	0.003	0.001
S+C+H+SC	0.019	0.033	0.029
S+C+H+SH	**0.161**	**0.272**	**0.241**
S+C+H+CH	0.025	0.043	0.038
S+C+H+SC+SH	**0.131**	0.021	**0.134**
S+C+H+SC+CH	0.018	0.003	0.018
S+C+H+SH+CH	**0.197**	0.032	**0.201**
S+C+H+SC+SH+CH	**0.184**	0.001	**0.128**
S+C+H+SC+SH+CH+SCH	**0.222**	0.000	**0.105**

Adapted from: Fletcher, D., Dillingham, P.W.: Model-averaged confidence intervals for factorial experiments. Comput. Stat. Data. An. **55**, 3041–3048, ©2011, with permission from Elsevier

Compared to both AIC and BIC weights, those obtained using AICc are concentrated on three of the smaller models, only one of which contains an interaction. The AIC and BIC weights are concentrated on five of the larger models, all of which contain at least one interaction; as $n = 16$, BIC gives slightly more weight to the smallest of these five models (Sect. 3.2.1).

3.3.4 Toxicity of a Pesticide

When we introduced the toxicity example in Sect. 1.3.4, we considered model weights based on AIC. We now compare these with weights obtained using AIC(w), stacking and bootstrap-based focussed model averaging.

For AIC(w) and stacking, we chose the weights that minimised (3.12) and maximised (3.13) respectively. As discussed in Sect. 1.3.4, it is natural to work on the $\log_2$-scale in this example, so model averaging was carried out on this scale, followed by back-transformation. For focussed model averaging we again used 10^4 bootstrap samples.

Table 3.5 Model weights for the toxicity experiment obtained using AIC, AIC(w), stacking and bootstrap-based focussed model averaging

| Model | AIC | AIC(w) | Stacking | Focussed model averaging | | | |
| | | | | $\pi_0 = 0.5$ | | $\pi_0 = 0.9$ | |
				Male	Female	Male	Female
1	0.004	0.077	0.020	0.027	0.038	0.093	0.000
2	0.235	0.075	0.091	0.000	0.962	0.000	0.458
3	0.553	0.849	0.889	0.909	0.000	0.907	0.531
4	0.209	0.000	0.000	0.063	0.000	0.000	0.011

The model weights for each of the five methods are shown in Table 3.5. For AIC(w), the correlations between the $\widehat{\eta}_{mi}$ in (3.11) are all at least 0.95, with that for models 3 and 4 being almost one. Thus models 3 and 4 are virtually indistinguishable, in terms of estimating the η_i, and model 4 is given zero weight. For stacking, the correlations between the $\widehat{\eta}_{m[-i]}$ in (3.14) are smaller than those for $\widehat{\eta}_{mi}$, but still all at least 0.92, and that for models 3 and 4 is again almost one, leading to zero weight for model 4. In Sect. 3.7.2 we consider the effect on these weights of removing the simplex-constraint.

For all the methods models 1 and 4 receive little weight, except for AIC, which gives some weight to model 4. AIC, AIC(w) and stacking all give most weight to model 3, with the AIC(w) and stacking weights being very similar. For males the focussed weights are similar to those based on AIC(w) and stacking, while for females they differ from those based on the other three methods, especially when $\pi_0 = 0.5$, where model 2 is given almost all the weight.

The model-averaged estimates of the required dose-level are shown in Table 3.6. Each of these is obtained by back-transformation of the estimate for x_0. As suggested in Sect. 1.3.4, when $\pi_0 = 0.5$ the model-averaged estimate appears to be robust to the choice of weights; differences between the estimates are greater when $\pi_0 = 0.9$, especially for females.

Table 3.6 Model-averaged estimates for the dose-levels (μg) of trans-cypermethrin that lead to 50 and 90% of individuals being affected, separately for each sex, obtained using AIC, AIC(w), stacking and bootstrap-based focussed model averaging

Probability affected	Sex	AIC	AIC(w)	Stacking	Focussed
0.5	Male	4.7	4.9	4.8	4.8
	Female	9.8	9.5	9.7	9.5
0.9	Male	16.5	16.7	16.1	16.6
	Female	50.7	51.5	52.9	47.7

3.4 Interval Estimation

Although it is helpful to have a good method for calculating a model-averaged point estimate, it is often more important to be able to produce a reliable model-averaged confidence interval. Indeed, one of the main reasons for using model averaging is to allow for model uncertainty when calculating such an interval. Model selection typically leads to an interval which has lower coverage, and is less stable, than a model-averaged interval [55, 131].

As the weights used to calculate $\widehat{\theta}$ in (3.1) are random variables, and the estimates of θ from different models may be correlated, calculation of a reliable model-averaged confidence interval is challenging [88, 166].

When model M is assumed to be true, an interval based on this model should have good coverage properties, at least asymptotically. However, if some of the elements of β_M are small, this interval may be substantially wider than a model-averaged interval. Conversely, a model-averaged interval is likely to have a coverage rate that is lower than the nominal level, as it will generally be narrower than an interval from model M. This trade-off between coverage and interval-width was evident in the simulation results discussed in Sect. 1.3.3.

We need to make a somewhat arbitrary decision about the reduction in coverage we are prepared to accept for a specified reduction in interval width. Even if we quantify this trade-off, it will usually be difficult to assess the properties of a model-averaged confidence interval analytically, except in special cases [106, 107, 109]. These issues make good interval estimation in FMA problematic, and there is a need for more research in this area.

We consider three types of model-averaged confidence interval:

1. Wald interval
2. Percentile-bootstrap (PB) interval
3. Model-averaged tail area (MATA) interval

Both the Wald and MATA interval can be used with any of the methods for determining model weights, such as AIC, AIC(w), stacking, or focussed weights. Calculation of the PB interval is analogous to the use of bagging for point estimation, and can be used in conjunction with any method of model selection, such as AIC.

3.4.1 Wald Interval

Suppose we wish to construct a $100\,(1 - 2\alpha)\,\%$ confidence interval for θ. If we make the assumption that

$$T_{\widehat{\theta}} = \frac{\widehat{\theta} - \theta}{s}$$

has a N (0, 1) distribution, where s is an estimate of the standard error of $\widehat{\theta}$, we can use the well-known Wald interval, given by

$$\widehat{\theta} \pm zs, \tag{3.17}$$

where z is the $100\,(1-\alpha)$th percentile of the N (0, 1) distribution. Use of the Wald interval has the following drawbacks:

1. The distribution of $\widehat{\theta}$ may not be normal, even if each $\widehat{\theta}_m$ has a normal distribution, due to the model weights being random variables;

2. Estimation of the standard error of $\widehat{\theta}$ is difficult, again due to randomness in the model weights and because $\widehat{\theta}_{m_1}$ and $\widehat{\theta}_{m_2}$ will often be correlated ($m_1 \neq m_2$);

3. The uncertainty associated with the estimate of the standard error is not allowed for, which might be important if n is small;

4. There may be correlation between $\widehat{\theta}$ and s, which can lead to skewness in the distribution of $T_{\widehat{\theta}}$.

In order to derive an expression for the standard error of $\widehat{\theta}$, [21] adopted a framework in which the models are regarded as a random sample from a population of models, with expectations being taken over this population, as well as conditional upon a single model.[10] In addition, they assumed that the bias associated with a model has an expectation of zero over the population of models, which means that $\widehat{\theta}$ is unbiased.[11] For mathematical convenience they also assumed that the weights are fixed, as in focussed model averaged point estimation (Sect. 3.2.3), and that the estimates from any two models have a perfect (positive) correlation. This leads to an estimate of the standard error of $\widehat{\theta}$ being given by

$$s = \sum_{m=1}^{M} w_m \left\{ \widehat{b}_m^2 + \widehat{v}_m \right\}^{1/2}, \tag{3.18}$$

where $\widehat{b}_m$ and $\widehat{v}_m$ are estimates of

$$b_m = \mathrm{E}\left(\widehat{\theta}_m \middle| m\right) - \theta$$

and

$$v_m = \mathrm{E}\left[\left\{ \widehat{\theta}_m - \mathrm{E}\left(\widehat{\theta}_m \middle| m\right) \right\}^2 \middle| m \right].$$

---

[10] An alternative derivation avoids the notion of selecting a random sample from a population of models [24]. However this involves regarding θ as a weighted mean of least-false values of θ.

[11] It has been wrongly claimed that $\widehat{\theta}$ is often assumed to be unbiased [48]. The only theory that involves this assumption (asymptotically) is the local misspecification framework (Sect. 3.2.3).

We can obtain $\widehat{v}_m$ in the usual way after fitting model m, and use of

$$\widehat{b}_m = \widehat{\theta}_m - \widehat{\theta} \tag{3.19}$$

is motivated by the assumption that $\widehat{\theta}$ is unbiased.[12]

Given the assumptions underlying use of the estimate in (3.18), it is natural to consider alternatives.[13] One is motivated by analogy with the square root of the model-averaged posterior variance in (2.5), and is given by [24, 25]

$$s = \left[\sum_{m=1}^{M} w_m \left\{\widehat{b}_m^2 + \widehat{v}_m\right\}\right]^{1/2}. \tag{3.20}$$

The Cauchy-Schwartz inequality implies that this interval is wider than the one based on (3.18). As both of these intervals are centred at $\widehat{\theta}$, the one based on (3.20) will have a higher coverage rate. Note that the form of s in either (3.18) or (3.20) suggests that $\widehat{\theta}$ and s will be correlated. As discussed above, this is one of the potential disadvantages of the Wald interval.

In order to make some allowance for the uncertainty in s (point 3 above), [24] proposed the following heuristic alternative to (3.18)

$$s = \sum_{m=1}^{M} w_m \left\{\widehat{b}_m^2 + \left(\frac{t_m}{z}\right)\widehat{v}_m\right\}^{1/2},$$

where t_m is the $100(1-\alpha)$th percentile of the t distribution with degrees of freedom given by the residual degrees of freedom for model m. A similar adjustment to (3.20) leads to

$$s = \left[\sum_{m=1}^{M} w_m \left\{\widehat{b}_m^2 + \left(\frac{t_m}{z}\right)\widehat{v}_m\right\}\right]^{1/2}. \tag{3.21}$$

Both [93] and [142] derive expressions for the MSE of $\widehat{\theta}$, in the context of focussed model averaging (Sect. 3.2.3), based on the assumption that model M is true.[14] It would be interesting to use their results to derive a Wald interval, as they provide asymptotic approximations to the correlation between the estimates from two models, rather than assume that these estimates are perfectly correlated.

The Wald interval given by (3.17) and (3.18) was criticised by [88], who showed that it could have poor asymptotic coverage. They proposed an alternative Wald-like interval that has the same width as a Wald interval from model M, and therefore

---

[12]Even if $\widehat{b}_m$ is unbiased, $\widehat{b}_m^2$ will be biased as an estimate of b_m^2, but analytical bias-adjustment would involve estimation of the correlation between $\widehat{\theta}_{m_1}$ and $\widehat{\theta}_{m_2}$ $(m_1 \neq m_2)$, and any decrease in bias might be offset by an increase in variance.

[13]There is also a logical problem associated with use of (3.18) [24].

[14]Both [21] and [24] wanted to avoid assuming that the true model is in the model set.

provides no obvious advantages over simply using the Wald interval from model M [33, 37, 38, 105, 128, 192, 193, 209, 216].

In order to avoid the simplifying assumptions used by [21, 27] considered a bootstrap-based estimate of s, in the context of linear models. They estimated s by the standard deviation of the non-parametric bootstrap-sampling distribution of $\widehat{\theta}$, with the model weights being based on an information criterion. Weights based on AIC, rather than BIC or AICc, were found to provide a better estimate of s. This approach is still prone to problems if the sampling distribution of $\widehat{\theta}$ is not close to normal.

3.4.2 Percentile-Bootstrap Interval

If we use bagging (Sect. 3.2.2) to obtain a point estimate, it is also natural to define a $100\,(1-2\alpha)\%$ confidence interval as the 100αth and $100\,(1-\alpha)$th percentiles of the bootstrap-sampling distribution of $\widehat{\theta}_{(b)}$ in (3.8) [21, 55]. We refer to this as the percentile-bootstrap (PB) interval. We focus on the case where the bootstrap samples are generated from model M, but one could use a non-parametric bootstrap.

Other bootstrap-based approaches to calculating a model-averaged confidence interval have been proposed. A weighted percentile-bootstrap approach was considered by [21], but the weighting used does not appear to provide any obvious benefit. Likewise, [55] proposed an estimate of the standard error of the bagging estimate in (3.8) that can be used in a Wald-based interval.[15] As with any Wald-based interval, this is likely to perform poorly if the distribution of the bagging estimate is non-normal. Interestingly, for the special case of two normal linear models, [108] showed that this interval is again no better than the Wald interval from model M. In the examples that follow, the bootstrap-based method we consider is the simple PB interval.

3.4.3 MATA Interval

Given the difficulties in assessing the sampling distribution of $\widehat{\theta}$, [60] proposed an interval that is akin to a model-averaged credible interval in BMA. By analogy with (2.6), the interval is chosen by setting the model-averaged lower and upper error rates equal to the required error rates. The lower limit of the $100\,(1-2\alpha)\,\%$ confidence interval is the value of θ_0 that satisfies the equation

$$\sum_{m=1}^{M} w_m \alpha_m\,(\theta_0) = \alpha, \tag{3.22}$$

---

[15]This estimate is not simply the standard deviation of the $\widehat{\theta}_{(b)}$ in (3.8) [21].

where $\alpha_m(\theta_0)$ is an estimate of the error rate associated with using θ_0 as a lower $100(1 - 2\alpha)\%$ confidence limit for θ when model m is true. The upper limit is defined in exactly the same way, with $\alpha_m(\theta_0)$ being an estimate of the upper error rate. The name for this interval arises from the fact that $\alpha_m(\theta_0)$ is the tail area of an appropriate sampling distribution [181].[16]

There are two advantages in using this interval. First, we avoid the need to specify the sampling distribution for $\widehat{\theta}$. Second, the method used to calculate $\alpha_m(\theta_0)$ can be tailored to the setting, in that we can choose a method that leads to a reliable confidence interval when based on the true model. For example, if we are averaging over a set of normal linear models, we would expect a t-interval based on the true model to work well, which suggests calculating $\alpha_m(\theta_0)$ using a t-distribution with degrees of freedom equal to the residual degrees of freedom for model m [181]. Likewise, when n is large and we are averaging over a set of binomial or Poisson models, we can calculate $\alpha_m(\theta_0)$ using a standard normal distribution. If a profile likelihood interval in the single model setting is likely to work well, we would calculate $\alpha_m(\theta_0)$ using a χ^2 approximation to the sampling distribution of the likelihood ratio test statistic under model m [60].

In order to to make clear which method is used to calculate $\alpha_m(\theta_0)$, we adopt the following notation: MATA-T, MATA-W and MATA-P refer to intervals based on the t-interval, Wald interval and profile likelihood interval respectively. For some of the examples in Sect. 3.5, we make use MATA-W intervals; examples of MATA-P and MATA-T intervals are given in [60, 181] respectively. Note that the MATA-W interval will generally be asymmetric around $\widehat{\theta}$, unlike the Wald interval in (3.17).

Although use of (3.22) does not guarantee that the resulting interval achieves the required error rates, the similarity of (3.22) to the definition of the model-averaged credible interval in (2.6) suggests it could perform well in many situations. This analogy was part of the motivation for development of the MATA-P interval [60], a slightly different justification being used by [181] for the MATA-W and MATA-T intervals. Neither of these arguments involve a theoretical assessment of the resulting error rates, but the simulation studies reported in [60, 107, 109, 181] indicate the potential for the MATA interval to perform well relative to the Wald interval in (3.17).

As pointed out by [60], the MATA interval will clearly inherit the strengths and weaknesses of the interval upon which it is based. For example, we would not consider using a MATA-P interval if a profile likelihood interval based on the true model is unlikely to work well. This self-evident result is borne out by the conclusions of [106], who derive expressions for the coverage and width of a MATA-P interval in a simple normal linear model setting. They find that this interval performs poorly

---

[16]It has been wrongly claimed that use of this interval involves assuming that the largest model is not in the model set [48].

when a profile likelihood interval is likely to have poor coverage, such as when we have many parameters to maximise over when calculating the profile likelihood.[17]

For the same linear model setting, [107] used analytical expressions to assess the properties of the MATA-W interval, and found that it can perform well if the weight given to the larger of the two models is as high as possible amongst the weighting schemes being considered. If we allow any choice of weights that lie on the unit simplex, this implies that we should simply use the interval from model M (Sect. 3.4.1). It also implies that a MATA-W interval based on AIC weights will be preferable to one based on AICc weights, and to one based on BIC weights (unless $n < 8$). Further theoretical and simulation work is needed on the properties of MATA intervals for a range of settings.

The MATA interval has the following connection with the PB interval described in Sect. 3.4.2. The lower limit of the PB interval can be shown to satisfy (3.22) if, in the notation of Sect. 3.2.2, we set $w_m = B_m/B$ and $\alpha_m(\theta_0)$ is the proportion of times that $\widehat{\theta}_m < \theta_0$ in the B_m bootstrap samples for which model m is selected as the best. A similar result applies to the upper limit. This suggests a potential disadvantage of the PB interval, compared to the MATA interval, in that this version of $\alpha_m(\theta_0)$ will not be a precise estimate of

$$p\left(\widehat{\theta}_m < \theta_0 \middle| S = m\right)$$

when B_m is small, where S is a random variable denoting the model selected as the best. This disadvantage will be ameliorated somewhat by the fact that the weight for model m will be small when B_m is small.[18]

A studentised-bootstrap version of the MATA interval (MATA-SB) was proposed by [214]. This uses a parametric studentised-bootstrap approach to estimate $\alpha_m(\theta_0)$ in (3.22), where the bootstrap samples are generated from the fitted version of model M. As for the other MATA intervals, we would expect the MATA-SB interval to work well in a setting where a parametric studentised-bootstrap interval performs well for the true model. In particular, it should perform better than the PB interval, which involves no studentisation, and at least as well as MATA-T if the number of bootstrap samples is large enough. The MATA-SB interval provides a useful alternative to the MATA-T, MATA-W and MATA-P intervals when the assumptions underlying these intervals are not met (Sect. 3.4.3). More work is needed to assess the properties of the MATA-SB interval in a range of settings.

---

[17] Unfortunately, the work of [106] has led to the impression that the MATA interval will not perform well in general [48].

[18] A similar issue arises when using a technique such as RJMCMC in the Bayesian setting (Sect. 2.2.1), where a large number of iterations may be required in order to visit each model often enough to obtain reliable estimates of both the posterior model probabilities and the posteriors for those parameters in models with low posterior model probabilities.

3.5 Examples

3.5.1 Sea Lion Bycatch

Table 3.7 shows model-averaged 95% confidence intervals for the total sea lion bycatch in each fishery, together with a 95% Wald interval for each model. The MATA-W interval is based on AIC weights, and was first seen in Table 1.3, while the PB interval is based on using AIC for model selection in each of 10^4 bootstrap samples. The single-model intervals and the MATA-W interval are obtained by back-transformation of the corresponding interval for $\log \mu_i$ ($i = 1, 2, 3$). Although both types of model-averaged interval use AIC (for model weights or for model selection) they lead to quite different confidence limits.

3.5.2 Ecklonia Density

Table 3.8 shows four MATA-W confidence intervals for the mean density of ecklonia in each zone, together with the Wald intervals from each model. Each interval is obtained by back-transformation of the corresponding interval for $\log \mu_i$ ($i = 1, 2, 3$). The weights used by each method are in Table 3.2 of Sect. 3.3.2.

As with the point estimates (Sect. 3.3.2), the differences between the model-averaged intervals are relatively small for zones 1 and 3. For zone 2, however, the interval based on focussed weights is markedly narrower, as it gives almost all the weight to the smaller model.

Table 3.7 95% confidence intervals for total sea lion bycatch (to the nearest integer) obtained from each of two models and by model averaging. The model-averaged intervals are the MATA-W interval (based on AIC weights) and the PB interval (based on using AIC for model selection)

Species		Lower	Upper
Scampi	Model 1	22	56
	Model 2	19	181
	MATA-W	21	118
	PB	0	136
Squid	Model 1	74	192
	Model 2	61	180
	MATA-W	69	190
	PB	56	168
Other	Model 1	3	7
	Model 2	2	74
	MATA-W	2	35
	PB	0	31

Table 3.8 95% confidence intervals for the mean density of ecklonia (individuals per quadrat) in three zones, for each model and three versions of the MATA-W interval, using weights determined by AIC, AIC(w), stacking or bootstrap-based focussed model averaging

Zone		Model 1	Model 2	Model-averaged intervals using MATA-W			
				AIC	AIC(w)	Stacking	Focussed
1	Lower	24.1	9.5	9.9	10.0	9.8	9.8
	Upper	46.6	30.0	42.0	42.4	40.8	40.4
2	Lower	24.1	17.6	18.5	18.6	18.2	23.8
	Upper	46.6	64.0	61.1	60.8	62.0	47.2
3	Lower	24.1	28.0	25.9	25.7	26.2	26.4
	Upper	46.6	73.2	70.7	70.3	71.4	71.7

Table 3.9 Mean coverage rate of a 95% Wald interval for a treatment combination mean, averaged over the eight combinations

Scenario	Best model			Model-averaged		
	AIC	AICc	BIC	AIC	AICc	BIC
Low	0.82	0.82	0.81	0.95	0.95	0.95
Medium	0.91	0.79	0.89	0.94	0.90	0.93
High	0.94	0.86	0.93	0.94	0.92	0.94

Adapted from: Fletcher, D., Dillingham, P.W.: Model-averaged confidence intervals for factorial experiments. Comput. Stat. Data. An. **55**, 3041–3048, ©2011, with permission from Elsevier

3.5.3 *Water-Uptake in Amphibia*

Simulation results from [59] are given in Tables 3.9 and 3.10, for an experiment involving two replicates. These show the coverage rates and relative widths of 95% Wald intervals for a treatment combination mean, averaged over the eight combinations. Each model-averaged Wald interval was calculated as in (3.17), using the estimate of s in (3.21).

As in Sect. 1.3.3, the results are given for three scenarios, corresponding to the true main effects and interactions being low, medium or high relative to the error variance. Of the three criteria used for model averaging, it appears that AIC performs best and AICc worst. This conclusion was also evident from the full set of simulation results described in [59].

These simulation results are in accord with theoretical results for the MATA-W interval given by [107] (Sect. 3.4.3). They found that AIC weights were preferable to AICc or BIC weights, in terms of coverage and interval-width, in a setting involving two normal linear models (Sect. 3.4.3).[19]

---

[19]This example also provides evidence that the Wald interval can perform well, despite the issues raised in Sect. 3.4.1.

Table 3.10 Mean width of a 95% Wald interval for a treatment combination mean relative to that for the full model, averaged over the eight combinations

Scenario	Best model			Model-averaged		
	AIC	AICc	BIC	AIC	AICc	BIC
Low	0.52	0.40	0.44	0.70	0.56	0.63
Medium	0.81	0.75	0.78	0.89	0.90	0.88
High	0.95	0.97	0.93	0.98	1.15	0.98

Adapted from: Fletcher, D., Dillingham, P.W.: Model-averaged confidence intervals for factorial experiments. Comput. Stat. Data. An. **55**, 3041–3048, ©2011, with permission from Elsevier

Table 3.11 95% MATA-W confidence intervals for the dose-levels (μg) of trans-cypermethrin that lead to 50% and 90% of individuals being affected, separately for each sex. The weights used to calculate each interval were determined by AIC, AIC(w), stacking, or bootstrap-based focussed model averaging

Probability affected	Sex		AIC	AIC(w)	Stacking	Focussed
0.5	Male	Lower	3.6	3.7	3.7	3.7
		Upper	6.3	7.1	6.4	6.5
	Female	Lower	6.9	6.2	6.7	6.6
		Upper	13.8	13.9	13.9	13.1
0.9	Male	Lower	10.2	10.1	10.1	10.1
		Upper	27.5	33.9	27.6	35.0
	Female	Lower	25.6	24.5	25.9	25.5
		Upper	107.9	109.5	110.3	101.7

3.5.4 Toxicity of a Pesticide

Table 3.11 shows 95% MATA-W intervals for the required dose-level, separately for each probability and sex. As with the point estimates in Sect. 3.3.4, these are based on weights obtained from AIC, AIC(w), stacking, and bootstrap-based focussed model averaging, and involve back-transformation of the corresponding interval for $\log_2$-dose. As with the model-averaged point estimates for this example, the choice of weights has little effect on the intervals when $\pi_0 = 0.5$, as would be expected from the discussion in Sect. 3.3.4. The clearest differences between the methods are for the upper limit when $\pi_0 = 0.9$.

3.6 Discussion

3.6.1 Choice of Scale

As discussed in Sect. 3.2, an issue that arises with point estimation in FMA is the choice of scale on which to perform the averaging. This issue does not arise in BMA, as the model-averaged posterior distribution is transformation-invariant (Sect. 2.5).[20]

We do not agree with authors who have argued that model-averaged point estimation should always take place on the original scale [26]. For example, when averaging over a set of GLMs, it is natural to calculate the arithmetic weighted mean in (3.1) on the linear-predictor scale, as the effects of the predictor variables are assumed be additive on this scale (Sect. 3.2). A similar argument arises in the context of analysing an experiment involving several treatments: if we need to transform the response variable in order to better satisfy the assumptions of a normal linear model, calculating a treatment mean makes most sense on this transformed scale [141].

For interval estimation, the Wald, MATA-W and MATA-T intervals are also not invariant to transformation, and we therefore need to consider the scale on which these are calculated. The MATA-P and PB intervals are both transformation-invariant, as likelihoods and percentiles are transformation-invariant.

3.6.2 Choice of Model Set

For simplicity, the examples we have considered all involve a relatively small model set. In some situations, there will be many possible predictors. A screening procedure may then be helpful, both for computational reasons and in order to reduce uncertainty in the estimation of the model weights. One approach advocated by [24] is careful selection of the models, driven by a clear idea as to their scientific merit. This is a laudable aim, but might be difficult to achieve in some settings, and we may still wish to have an alternative means of reducing the number of candidate models.

A simple screening procedure that has been developed in the context of high-dimensional regression models, involves assessing the strength of the relationship between each individual predictor and the response variable, in order to rank the predictors. This ranking is used to form $M + 1$ groups, with the first being the predictors with the strongest relationships and the last being those with the weakest relationships. After discarding the last group, model m is chosen to contain all the predictors in group m [10, 11, 65]. An alternative approach, which avoids using the original data to both screen the variables and to determine the model weights, involves sample-splitting during the screening procedure [65, 123]. In a similar vein, focussed model averaging based on the full set of singleton models (those with only one predictor) can lead to the same prediction performance as a larger set of nested

---

[20]Unless we use DIC weights, which can depend on the paramtetrisation (Sect. 2.5).

models [32, 88]. This is a topic worthy of further research, as it has the potential to greatly simplify the choice of model set and to reduce the computational effort.

The weighted-average least squares (WALS) approach to model averaging uses a quasi-Bayesian approach to combining frequentist point estimates [136]. An interesting feature of this method is the orthogonalisation of predictor variables, which reduces the number of main effects models from 2^k to k, where k is the number of predictor variables, an approach that has also been advocated by [8] when using AIC(w) [79] (Sect. 3.2.3). This reduction of the number of models is related to the use of singleton models discussed above. Extensions of WALS to other settings have been proposed, including GLMs [46] and linear models with errors that are dependent and/or heteroscedastic [135]. The latter has connections with a generalised least squares version of AIC(w) [129].

A screening procedure proposed by [212] is based on the values of AIC_m or BIC_m for half the data, the other half being used to determine the model weights. Fence methods, which are used in model selection, provide an alternative means of reducing the model set, although they can be computationally intensive [101, 144]. In the context of logistic regression with 15 possible predictor variables, [36] chose the model set to be those models that were selected at each step of a forward-search procedure. A bootstrap-based approach to screening predictor variables, rather than models, was proposed by [13]. This has the advantage of reducing the cost of future studies if some predictor variables are expensive to measure [20, 49].

3.6.3 Confidence Intervals

When comparing methods for calculating a confidence interval, it is common practice to first consider the coverage rate and then compare those methods which achieve good coverage by the mean width of the interval [59]. This has the disadvantage that what we mean by a good coverage rate is somewhat arbitrary. For example, in the model-averaging context, would we prefer an interval with 94% coverage and mean width equal to 80% of that for model M or one with 93% coverage and mean width equal to 60% of that for model M? Focussing primarily on coverage can lead to the idea that nothing can improve upon the largest model [48]. This can be misleading, as illustrated by the discussion of the water-uptake experiment in Sects. 1.3.3 and 3.5.3.

An alternative approach would be to choose the interval that has the smallest width, after the nominal confidence level associated with that interval has been adjusted to ensure that it actually achieves the required coverage rate [150]. For example, if simulations suggest that a 95% model-averaged confidence interval has a true coverage rate of 92%, we could increase the nominal confidence level until the interval achieves a true coverage rate of 95%. Having calibrated the intervals in this way, we could choose the one with the smallest mean width. This might be a useful basis for comparison, even if we did not expect to make use of such a calibration when applying the method.[21] There is scope for further research in this area.

A variation on the MATA interval has been proposed by [210]. This is slightly simpler to calculate, but no longer has the appealing property of satisfying (3.22). Likewise, in the context of using vector autoregression models in macro-economics, [122] provide heuristic arguments for a model-averaged interval that has endpoints which are functions of the endpoints of each single-model interval.

In machine learning, calculation of confidence intervals is a more recent innovation [188], as researchers have typically been satisfied with a cross-validation-based estimate of the prediction error to be expected on new data.

3.6.4 Mixed Models

Two mixed-model versions of AIC have been proposed for model selection. If we are interested in population-level inference, the random effects can simply be regarded as a device for modelling the covariance structure. We can then use marginal AIC (mAIC), based on (3.3) with p_m equal to the total number of parameters, including the variance components [144, 184]. Unlike the behaviour of AIC in fixed effects models, use of mAIC can lead to overly-simple models being selected if some of the models have different covariance structures [76, 144]. The calculation of mAIC can also be based on restricted maximum likelihood (REML) estimates of the variance components, which are less biased than those based on maximum likelihood [38, 144].

If we are interested in predictions for a specific level of one or more random effects, a conditional version of AIC (cAIC) can be used to compare models with different variance parameters [184]. This criterion is based on a conditional likelihood, as it involves the distribution of y given the random effects in the study. There have been several suggestions as to a suitable correction for overfitting in cAIC, reflecting the fact that a measure of complexity in this setting is not clear-cut [144].

The work of [184] on cAIC was extended by [120], in order to avoid specification of the form of the covariance matrix for the random effects; see also [161]. Recently, [211] have proposed a conditional generalised information criterion, derived without needing to assume that the true model is in the model set. For the generalised linear mixed model (GLMM) setting, [47, 208] proposed slightly different versions of cAIC. The former makes use of a conditional profile likelihood [37, 201] and a correction for overfitting bootstrap-based, akin to that proposed for mAIC by [169].

Use of mAIC or cAIC in (3.2) provide mAIC or cAIC weights [33]. An alternative to cAIC weights are those based on a conditional version of AIC(w) developed for

---

[21]Conversely, we could adjust the nominal confidence level for each interval until they all have the same width, and then choose the one with the highest true coverage rate [150].

GLMMs by [223], and simulations have shown that the latter can perform better. Mixed-model versions of stacking would also be possible, with theory for model selection in the linear mixed model setting [58] suggesting that these will be asymptotically equivalent to mAIC (or cAIC) weights.

As in model selection, point estimation will be more sensitive to reliable specification of the mean structure than the covariance structure, whereas good interval estimation will typically depend on both. An extra complication arises if some of the variance parameters are close to zero, as this can lead to the computational and theoretical issues associated with a parameter being close to a boundary [144].

3.6.5 Missing Data

Two approaches to model averaging in the presence of missing data were considered by [165]. One involves adjustment of the weights, while the other uses any method of FMA on a single imputed set of data. In a simulation study involving binary logistic regression, they found that the second approach performs better. Use of multiple, rather than single, imputation was suggested by [166]. An approach that avoids data imputation was proposed by [219], and is based on AIC(w) [79] (Sect. 3.2.3). A modified version of the missing-data indicator approach of [160] was considered by [42, 43]. In the ecological setting, [148] discuss the issues that arise when using multiple imputation with model averaging, emphasising the extra problems that can occur when the data are missing not at random. In the context of analysing time-to-event data from a cross-over trial, [203] considered the use of model averaging in conjunction with multiple imputation.

3.6.6 Summing Model Weights

By analogy with the use of a PIP in classical BMA (Sect. 2.2), [24] suggested calculating the sum of the weights across all models that contain a specific predictor variable, in order to obtain a measure of the relative importance of that variable. As with the use of a PIP, this is less useful than a comparison of model-averaged estimates of the expected response for suitably-chosen values of the predictor variables (Sect. 1.4). Care is also needed in interpreting these summed weights, as they can be sensitive to the choice of model set [146]. Potential problems with summed weights are discussed by [26, 48, 68, 146, 173]. In particular, [68] provide a clear rebuttal of the defence of summed weights in [74].

3.7 Related Literature

3.7.1 Information-Criterion Methods

Two small-sample modifications to AIC other than AICc have been suggested for
normal linear models. These involve slightly different modifications to the maximum
likelihood estimate of the error variance, and use the same correction for overfitting
as in AICc [139] or AIC [38]. They appear not to have been used much in the scientific
literature. An AIC-like weight based on likelihood-based cross validation has been
considered by several authors [48, 85, 185, 199]. However, this approach is less
direct than stacking (Sect. 3.2.3).

In the context of model selection, [52, 54] showed that a range of selection criteria,
including AIC and Mallows' C_p, could be expressed in terms of a covariance penalty,
which has links with the notion of generalised degrees of freedom [85, 170, 207,
213]. These ideas were transferred to the model averaging context by [171]. Recently,
[134] have proposed generalised versions of AIC and BIC for use with misspecified
generalised linear models, their primary focus being high-dimensional models.

For model selection in ecology, [19] proposed that information criteria be compared using simulations in which the true model differed slightly from one simulation
run to another, in a way that reflected how the true data-generating mechanism might
change if we were to repeat a study. They used simulation to show that a high level of
heterogeneity between replicate datasets may lead to BIC having a better prediction
performance than both AIC and AICc. Their conclusions suggest that in practice it
may be difficult to provide clear recommendations as to which of these criteria is to
be preferred, as the amount of variation between replicate datasets will often be hard
to quantify.

A novel use of cross validation can be found in [222], who suggested using it to
determine the best choice of model-selection procedure (e.g. AIC or BIC) for a given
dataset. It would be interesting to see how well such a two-stage approach might help
determine the best method for obtaining a model-averaged estimate or confidence
interval.

3.7.2 Constraints on Optimal Weights

Although it may seem natural to constrain the model weights to lie on the unit simplex,
particularly if the estimates from some models are highly correlated (Sect. 3.2.3),
several authors have considered arguments for relaxing this constraint.

In the context of using stacking for high-dimensional linear models and GLMs,
[10, 11] argued that the simplex-constraint can be too restrictive (Sect. 3.6.2), and
proposed instead that the weights should lie in the unit hypercube, i.e. $w_m \in [0, 1]$
for all m [32, 65, 123]. This situation arises because the screening procedure they

Table 3.12 Model weights for the toxicity experiment using AIC(w) and stacking with different weight-constraints (weights lie on the unit simplex, lie in the unit-hypercube, or sum to one). For AIC(w), the weights constrained to sum to one are highly unstable and therefore not shown

Model	AIC(w)			Stacking		
	Simplex	Hypercube	Sum to one	Simplex	Hypercube	Sum to one
1	0.077	0.050	–	0.020	0.000	−0.121
2	0.075	0.093	–	0.091	0.101	1.262
3	0.849	0.818	–	0.889	0.861	7.859
4	0.000	0.000	–	0.000	0.000	−7.999
Sum	1.000	0.961	–	1.000	0.963	1.000

use (Sect. 3.6.2) leads to a set of models that have no predictors in common, which means that the choice $w_m = 1$ (for all m) can be optimal.[22]

As mentioned in Sect. 1.6, several authors have considered allowing the weights to be negative, but still sum to one [6, 32, 94, 136, 180]. However, keeping the non-negativity constraint can help address issues with model-redundancy (Sect. 3.2.3).[23] For example, without this constraint two models that give very similar predictions can have large weights of opposite signs, which seems undesirable. This is akin to the issue that arises if two predictor variables in a regression model are strongly positively correlated, where shrinkage methods are sometimes used to constrain the regression coefficients [84].

The effect of the choice of constraints used in both AIC(w) and stacking is illustrated in Table 3.12. This shows the weights for each method using three types of constraint: the usual simplex-constraint, the unit-hypercube constraint, and the sum-to-one constraint, which allows the weights to be negative.

If we first consider AIC(w), the correlations between the $\widehat{\eta}_{mi}$ in (3.11) are all at least 0.95, with that for models 3 and 4 being almost one. Thus models 3 and 4 are virtually indistinguishable, in terms of estimating the η_i. As a result, the weights that are only constrained to sum to one are highly unstable, with many local optima, and therefore not shown in the table.

For stacking, the correlations between the $\widehat{\eta}_{m[-i]}$ in (3.14) are smaller than those for $\widehat{\eta}_{mi}$, but still all at least 0.92, and the correlation for models 3 and 4 is again almost one. In this case, however, all the weights are relatively stable. Allowing the weights to be negative leads to those for models 3 and 4 being of opposite sign.

The weights involving the simplex and unit-hypercube constraints are very similar, for both AIC(w) and stacking; see [15] for an argument as to why this will often be the case for stacking.

---

[22]This procedure is similar to use of all possible singleton models in the context of focussed model averaging (Sect. 3.6.2) [32]. In order for $\widehat{\theta}$ to be consistent, however, [32] require the weights to sum to one, as each $\widehat{\theta}_m$ is consistent [88].

[23]This constraint can also be useful for generalisation of the conclusions [15].

3.7.3 AIC(w)

In the context of normal linear models, a minor modification to the correction for overfitting in AIC(w) was proposed by [220]. The criterion proposed by [200] for selecting model weights is a model-averaging version of the Prediction Criterion [7] for model selection, and can have better finite-sample properties than AIC(w). A version of AIC(w) that involves a tuning parameter in the correction for overfitting was put forward by [227]. Use of AIC(w) when averaging over a set of semi-parametric varying-coefficient models has been considered by [118]; see also [119, 127]. The application of AIC(w) to forecast-combination has been considered by [80]. The asymptotic behaviour of AIC(w) under the local-misspecification assumption of [88] (Sect. 3.2.3) has been considered by [128].

3.7.4 Machine Learning Methods

Stacking is closely related to the concept of a super learner [151, 162, 186]. The asymptotic behaviour of stacking in the context of linear models, and under the local-misspecification assumption of [88] (Sect. 3.2.3), has been studied by [128]. Application of stacking to non-parametric models has been discussed by [183]. Use of stacking for density estimation [174] has close links with BSP (Sect. 2.3.2). Both bagging and stacking can be used in meta-learning, where a higher level of learning occurs through experience over several applications of one or more learners [115].

Two other model-averaging techniques commonly used in machine learning are boosting and random forests. Boosting involves generating a sequence of combined learners and using a resampling-based approach to repair weaknesses of the current combination [50, 63, 64, 84, 159, 163]. Random forests involves "growing" many regression or classification trees to randomised versions of the training data, and averaging them [18]. Boosting provides a means of bias-reduction, and sometimes of variance-reduction, while random forests achieves variance-reduction through the averaging process.

An idea that features in many ensemble learning methods is the desire to ensure that different models (learners) are based on quite different methods, such as linear discriminant analysis and random forests in classification [15]. This is related to the concept of discrete model averaging discussed by [49] (Sect. 1.4).

Ensemble averaging of neural networks has been considered by [147], and use of the dropout technique with such networks is closely related to model averaging [66]. A Bayesian approach to combining support-vector-machines has been described by [225]. Excellent introductions to machine learning methods are provided by [56, 84, 217].

3.7.5   *Focussed Methods*

Examples of contexts in which the local-misspecification assumption of [88] has been used are semi-parametric models [37], linear mixed models [33], multinomial models [191], generalised rank regression models [224] and structural equation models [102]. A variation on the work of [88] was considered by [121], the difference being a minor change in the estimate of the MSE of $\widehat{\theta}$.

In the context of linear models, [121] also considered a class of weights that includes those based on AIC and BIC. This class is specified by three parameters that are chosen to minimise an estimate of the average MSE of a set of estimates. They considered two types of set. The first were the estimates of the regression coefficients associated with those predictor variables that were the focus of the analysis. The second were the estimates of the expected value of the response variable for the observed values of all the predictor variables. They provided results from a simulation study which suggested that the optimal choice of weights outperformed those based on AIC(w), AIC and BIC.

The Focussed Information Criterion (FIC) was proposed by [35] as a focussed method of model selection. The value of FIC for model m (FIC$_m$) is based on an asymptotic approximation to the MSE of $\widehat{\theta}_m$, and an extension of this idea to model averaging was suggested by [88], who defined a model weight based on FIC$_m$ that takes the same form as the AIC weight in (3.2), together with a tuning parameter [36, 90, 194, 195, 215, 216, 218]. A similar approach was advocated by [117], in the context of linear models. As FIC$_m$ was developed for the purpose of model selection, FIC weights are not directly concerned with the properties of $\widehat{\theta}$. For model selection, [104] derived a version of FIC that does not require the local-misspecification assumption, in the context of comparing a set of parametric models to a non-parametric alternative.

3.7.6   *Miscellanea*

Many of the issues that arise in model averaging are applicable to forecast-combination, including potential problems caused by estimating the weights. For example, if the MSE of a combined forecast needs to be estimated, the resulting MSE will be larger than predicted (Sect. 3.2.3). In some cases, this can even lead to equal weights being preferable to estimated weights [1, 39, 71, 75, 179], a result that can also occur in model averaging [191]. Model averaging of predictive distributions has been considered in econometrics [28, 73, 77, 110, 189], and has links with prediction-based BMA (Sect. 2.3.2).

Conditions under which model-averaged estimates based on AIC weights, BIC weights, AIC(w) or stacking are consistent has been considered by [221] in the context of linear models.

Model averaging using the leave-one-out bootstrap was proposed by [157]. In the context of model selection, this version of the bootstrap can be less variable than leave-one-out cross validation, but usually requires a bias-adjustment that is difficult to determine [44, 53]. Interestingly, a number of other resampling approaches to model selection, including bootstrap-based estimation of the correction for overfitting in AIC, have not been transferred to the model averaging setting [30, 34, 54, 97, 98, 112, 172]. Use of the bootstrap to estimate prediction error was first considered in detail by [51].

The concept of a "model confidence set" was considered by [24]. One definition they proposed involves ranking the models in terms of their AIC weights and then determining the smallest set of consecutive models for which the sum of the weights is greater than or equal to the nominal "confidence level". In the context of model selection, [82] defined a such a set to be one that contains the best model with a specified level of "confidence".

A method for calculating a simultaneous confidence region for two or more parameters after model averaging has been proposed by [100]. This is a Wald region, and will only perform well in settings where a Wald interval for a single parameter is reliable (Sect. 3.4).

The difficulty of analytically assessing the finite-sample properties of a model-averaged estimate has been emphasised by [153], while [113, 114] have considered similar issues for an estimate based on a best model. Likewise, [57] discuss the issues involved in constructing a confidence interval with the desired coverage rate when using a shrinkage method such as the lasso; see also [164] (Sect. 1.4).

An unusual version of model averaging was proposed by [149]. This uses ideas from social choice theory, combined with resampling, to generate plausible versions of the true model. It involves summarising the results of a comparison of the different models across a set of plausible versions of the true model.

For averaging over quantile regression models, [168] proposed a sample-splitting algorithm, [132] suggested stacking, and [202] used FIC-based model weights. Several cross-validation procedures were proposed by [69] for averaging over both longitudinal and time series models. In the context of averaging threshold models, [81] proposed using AIC(w) (Sect. 3.2.3), while [70] suggested generalised cross validation [41]. Methods for determining model weights when averaging instrumental-variable models have been considered by [138], while [116] discuss a two-stage approach for performing BMA on such models. In the context of non-parametric regression, model averaging over the choice of predictor variables, as well as the choice of kernel, bandwidth-selection method and local-polynomial order has been considered by [86].

3.7.7 Software

As with BMA, there are several packages available in R that can be used for FMA:

1. MATA gives MATA-W and MATA-T intervals for any type of model weight
2. AICcmodavg, glmulti and MuMIn provides estimates and Wald intervals
3. MAMI implements AIC(w), stacking and lasso averaging [164, 167]

4. `SuperLearner` and `subsemble` provide super learners (Sect. 3.7.4)
5. `gbm` and `randomForest` are for boosting and random forests respectively
6. `ada`, `adabag`, `caretEnsemble`, `ipred`, `mboost` and `party` are ensemble-learning packages
7. `MCS` can be used for the model confidence set approach of [82]

Specialist software is also available for scientists working on ground-water modelling [152].

References

1. Aiolfi, M., Capistran, C., Timmermann, A.: Forecast combinations. In: Clements, M.P., Hendry, D.F. (eds.) Oxford Handbook of Economic Forecasting. Oxford University Press (2010)
2. Akaike, H.: Information theory as an extension of the maximum likelihood principle. In: Petrov, B.N., Csaki, F. (eds.) Second International Symposium on Information Theory, pp. 267–281. Akademiai Kiado, Budapest (1973)
3. Akaike, H.: A new look at the statistical model identification. IEEE Trans. Autom. Control. **19**, 716–723 (1974)
4. Akaike, H.: A Bayesian analysis of the minimum AIC procedure. Ann. I. Stat. Math. **30**, 9–14 (1978)
5. Akaike, H.: A Bayesian extension of the minimum AIC procedure of autoregressive model fitting. Biometrika **66**, 237–242 (1979)
6. Aksu, C., Gunter, S.I.: An empirical analysis of the accuracy of SA, OLS, ERLS and NRLS combination forecasts. Int. J. Forecast. **8**, 27–43 (1992)
7. Amemiya, T.: Selection of regressors. Int. Econ. Rev. **21**, 331–354 (1980)
8. Amini, S.M., Parmeter, C.F.: Comparisons of model averaging techniques: assessing growth determinants. J. Appl. Econ. **27**, 870–876 (2012)
9. Anderson, D.R., Burnham, K.P., White, G.C.: AIC model selection in overdispersed capture-recapture data. Ecology **75**, 1780–1793 (1994)
10. Ando, T., Li, K.-C.: A model-averaging approach for high-dimensional regression. J. Am. Stat. Assoc. **109**, 254–265 (2014)
11. Ando, T., Li, K.-C.: A weight-relaxed model averaging approach for high-dimensional generalized linear models. Ann. Stat. **45**, 2654–2679 (2017)
12. Arlot, S., Celisse, A.: A survey of cross-validation procedures for model selection. Stat. Surv. **4**, 40–79 (2010)
13. Augustin, N., Sauerbrei, W., Schumacher, M.: The practical utility of incorporating model selection uncertainty into prognostic models for survival data. Stat. Model. **5**, 95–118 (2005)
14. Bozdogan, H.: Akaike's information criterion and recent developments in information complexity. J. Math. Psychol. **44**, 62–91 (2000)
15. Breiman, L.: Stacked regressions. Mach. Learn. **24**, 49–64 (1996)
16. Breiman, L.: Bagging predictors. Mach. Learn. **24**, 123–140 (1996)
17. Breiman, L.: Heuristics of instability and stabilization in model selection. Ann. Stat. **24**, 2350–2383 (1996)
18. Breiman, L.: Random forests. Mach. Learn. **45**, 5–32 (2001)
19. Brewer, M.J., Butler, A., Cooksley, S.L.: The relative performance of AIC, AICC and BIC in the presence of unobserved heterogeneity. Methods Ecol. Evol. **7**, 679–692 (2016)
20. Buchholz, A., Hollnder, N., Sauerbrei, W.: On properties of predictors derived with a two-step bootstrap model averaging approach—a simulation study in the linear regression model. Comput. Stat. Data Anal. **52**, 2778–2793 (2008)

21. Buckland, S.T., Burnham, K.P., Augustin, N.H.: Model selection: an integral part of inference. Biometrics **53**, 603–618 (1997)
22. Buckland, S.T., Burnham, K.P., Augustin, N.H.: Rejoinder to the Letter to the Editors from Wagenmakers, E.-J., Farrell, S., Ratcliff, R. Biometrics **60**, 283 (2004)
23. Burnham, K.P., Anderson, D.R., White, G.C.: Evaluation of the Kullback-Leibler discrepancy for model selection in open population capture-recapture models. Biometrical. J. **36**, 299–315 (1994)
24. Burnham, K.P., Anderson, D.R.: Model Selection and Multimodel Inference: A Practical Information-Theoretic Approach, 2nd edn. Springer (2002)
25. Burnham, K.P., Anderson, D.R.: Multimodel inference understanding AIC and BIC in model selection. Sociol. Method. Res. **33**, 261–304 (2004)
26. Cade, B.S.: Model averaging and muddled multimodel inferences. Ecology **96**, 2370–2382 (2015)
27. Candolo, C., Davison, A.C., Demtrio, C.G.B.: A note on model uncertainty in linear regression. J. R. Stat. Soc. D-Stat. **52**, 165–177 (2003)
28. Carney, M., Cunningham, P.: Calibrating probability density forecasts with multi-objective search. Technical Report TCD-CS-2006-07, Trinity College, Dublin (2006)
29. Cavanaugh, J.E.: Unifying the derivations for the Akaike and corrected Akaike information criteria. Stat. Probabil. Lett. **33**, 201–208 (1997)
30. Cavanaugh, J.E., Shumway, R.H.: A bootstrap variant of AIC for state-space model selection. Stat. Sin. **7**, 473–496 (1997)
31. Cavanaugh, J.E.: A large-sample model selection criterion based on Kullback's symmetric divergence. Stat. Probab. Lett. **42**, 333–343 (1999)
32. Charkhi, A., Claeskens, G., Hansen, B.E.: Minimum mean squared error model averaging in likelihood models. Stat. Sin. **26**, 809–840 (2016)
33. Chen, X., Zou, G., Zhang, X.: Frequentist model averaging for linear mixed-effects models. Front. Math. China **8**, 497–515 (2013)
34. Chung, H.-Y., Lee, K.-W., Koo, J.-Y.: A note on bootstrap model selection criterion. Stat. Probab. Lett. **26**, 35–41 (1996)
35. Claeskens, G., Hjort, N.L.: The focused information criterion. J. Am. Stat. Assoc. **98**, 900–916 (2003)
36. Claeskens, G., Croux, C., Kerckhoven, J.V.: Variable selection for logistic regression using a prediction-focused information criterion. Biometrics **62**, 972–979 (2006)
37. Claeskens, G., Carroll, R.J.: An asymptotic theory for model selection inference in general semiparametric problems. Biometrika **94**, 249–265 (2007)
38. Claeskens, G., Hjort, N.L.: Model Selection and Model Averaging. Cambridge University Press, Cambridge (2008)
39. Claeskens, G., Magnus, J.R., Vasnev, A.L., Wang, W.: The forecast combination puzzle: a simple theoretical explanation. J. Forecast. **32**, 754–762 (2016)
40. Clyde, M.: Model uncertainty and health effect studies for particulate matter. Environmetrics **11**, 745–763 (2000)
41. Craven, P., Wahba, G.: Smoothing noisy data with spline functions: estimating the correct degree of smoothing by the method of generalized cross-validation. Numer. Math. **31**, 377–403 (1979)
42. Dardanoni, V., Modica, S., Peracchi, F.: Regression with imputed covariates: a generalized missing-indicator approach. J. Econ. **162**, 362–368 (2011)
43. Dardanoni, V., de Luca, G., Modica, S., Peracchi, F.: Bayesian model averaging for generalized linear models with missing covariates. No. 1311. Einaudi Institute for Economics and Finance (2013)
44. Davison, A.C., Hinkley, D.V.: Bootstrap Methods and their Applications. Cambridge University Press, Cambridge (1997)
45. Debray, T.P.A., Koffijberg, H., Nieboer, D., Vergouwe, Y., Steyerbergb, E.W., Moonsa, K.G.M.: Meta-analysis and aggregation of multiple published prediction models. Stat. Med. **33**, 2341–2362 (2014)

46. De Luca, G., Magnus, J.R., Peracchi, F.: Weighted-average least squares estimation of generalized linear models. J. Econ. (2018). https://doi.org/10.1016/j.jeconom.2017.12.007
47. Donohue, M.C., Overholser, R., Xu, R., Vaida, F.: Conditional Akaike information under generalized linear and proportional hazards mixed models. Biometrika **98**, 685–700 (2011)
48. Dormann, C.F., Calabrese, J.M., GuilleraArroita, G., Matechou, E., Bahn, V., Bartoń, K., Beale, C.M., Ciuti, S., Elith, J., Gerstner, K., Guelat, J., Keil, P., LahozMonfort, J.J., Pollock, L.J., Reineking, B., Roberts, D.R., Schröder, B., Thuiller, W., Warton, D.I., Wintle, B.A., Wood, S.N., Wüest, R.O., Hartig, F.: Model averaging in ecology: a review of Bayesian, information-theoretic, and tactical approaches for predictive inference. Ecol. Monogr. (2018). https://doi.org/10.1002/ecm.1309
49. Draper, D.: Model uncertainty yes, discrete model averaging maybe. Stat. Sci. **14**, 405–409 (1999)
50. Drucker, H., Cortes, C., Jackel, L.D., LeCun, Y., Vapnik, V.: Boosting and other ensemble methods. Neural Comput. **6**, 1289–1301 (1994)
51. Efron, B.: Estimating the error rate of a prediction rule: improvement on cross-validation. J. Am. Stat. Assoc. **78**, 316–331 (1983)
52. Efron, B.: How biased is the apparent error rate of a prediction rule? J. Am. Stat. Assoc. **81**, 461–470 (1986)
53. Efron, B., Tibshirani, R.: Improvements on cross-validation: the 632+ bootstrap method. J. Am. Stat. Assoc. **92**, 548–560 (1997)
54. Efron, B.: The estimation of prediction error: covariance penalties and cross-validation. J. Am. Stat. Assoc. **99**, 619–632 (2004)
55. Efron, B.: Estimation and accuracy after model selection. J. Am. Stat. Assoc. **109**, 991–1007 (2014)
56. Efron, B., Hastie, T.: Computer Age Statistical Inference, vol. 5. Cambridge University Press (2016)
57. Ewald, K., Schneider, U.: Uniformly valid confidence sets based on the Lasso. Electron. J. Stat. **12**, 1358–1387 (2018)
58. Fang, Y.: Asymptotic equivalence between cross-validations and Akaike information criteria in mixed-effects models. J. Data Sci. **9**, 15–21 (2011)
59. Fletcher, D., Dillingham, P.W.: Model-averaged confidence intervals for factorial experiments. Comput. Stat. Data. An. **55**, 3041–3048 (2011)
60. Fletcher, D., Turek, D.: Model-averaged profile likelihood intervals. J. Agr. Biol. Environ. Stat. **17**, 38–51 (2011)
61. Fletcher, D.: Estimating overdispersion when fitting a generalized linear model to sparse data. Biometrika **99**, 230–237 (2011)
62. Foster, D.P., George, E.I.: The risk inflation criterion for multiple regression. Ann. Stat. **22**, 1947–1975 (1994)
63. Freund, Y.: Boosting a weak learning algorithm by majority. Inform. Comput. **121**, 256–285 (1995)
64. Freund, Y., Schapire, R.E.: A decision-theoretic generalization of online learning and an application to boosting. J. Comput. Syst. Sci. **55**, 119–139 (1997)
65. Fu, P., Pan, J.: A review on high-dimensional frequentist model averaging. Open. J. Sta. **8**, 513–518 (2018)
66. Gal, Y., Ghahramani, Z.: Dropout as a Bayesian approximation: representing model uncertainty in deep learning. In: International Conference on Machine Learning, pp. 1050–1059 (2016)
67. Galipaud, M., Gillingham, M.A.F., David, M., Dechaume-Moncharmont, F.-X.: Ecologists overestimate the importance of predictor variables in model averaging: a plea for cautious interpretations. Methods Ecol. Evol. **5**, 983–991 (2014)
68. Galipaud, M., Gillingham, M.A.F., DechaumeMoncharmont, F.-X.: A farewell to the sum of Akaike weights: the benefits of alternative metrics for variable importance estimations in model selection. Methods Ecol. Evol. **8**, 1668–1678 (2017)

69. Gao, Y., Zhang, X., Wang, S., Zou, G.: Model averaging based on leave-subject-out cross-validation. J. Econ. **192**, 139–151 (2016)
70. Gao, Y., Zhang, X., Wang, S., Chong, T.T., Zou, G. Frequentist model averaging for threshold models. Ann. I. Stat. Math. (2018). https://doi.org/10.1007/s10463-017-0642-9
71. Genre, V., Kenny, G., Meyler, A., Timmermann, A.: Combining expert forecasts: can anything beat the simple average? Int. J. Forecast. **29**, 108–121 (2013)
72. George, E., Foster, D.P.: Calibration and empirical Bayes variable selection. Biometrika **87**, 731–747 (2000)
73. Geweke, J., Amisano, G.: Optimal prediction pools. J. Econ. **164**, 130–141 (2011)
74. Giam, X., Olden, J.D.: Quantifying variable importance in a multimodel inference framework. Methods Ecol. Evol. **7**, 388–397 (2016)
75. Graefe, A., Kchenhoff, H., Stierle, V., Riedl, B.: Limitations of ensemble Bayesian model averaging for forecasting social science problems. Int. J. Forecast. **31**, 943–951 (2015)
76. Greven, S., Kneib, T.: On the behaviour of marginal and conditional AIC in linear mixed models. Biometrika **97**, 773–789 (2010)
77. Hall, S.G., Mitchell, J.: Combining density forecasts. Int. J. Forecast. **23**, 1–13 (2007)
78. Hansen, M.H., Yu, B.: Model selection and the principle of minimum description length. J. Am. Stat. Assoc. **96**, 746–774 (2001)
79. Hansen, B.E.: Least squares model averaging. Econometrica **75**, 1175–1189 (2007)
80. Hansen, B.E.: Least-squares forecast averaging. J. Econ. **146**, 342–350 (2008)
81. Hansen, B.E.: Averaging estimators for regressions with a possible structural break. Economet. Theor. **25**, 1498–1514 (2009)
82. Hansen, P.R., Lunde, A., Nason, J.M.: The model confidence set. Econometrica **79**, 453–497 (2011)
83. Hansen, B.E., Racine, J.S.: Jackknife model averaging. J. Econ. **167**, 38–46 (2012)
84. Hastie, T., Tibshirani, R., Friedman, J.J.H.: The Elements of Statistical Learning, vol. 1. Springer, New York (2001)
85. Hauenstein, S., Wood, S.N., Dormann, C.F.: Computing AIC for black-box models using generalized degrees of freedom: a comparison with cross-validation. Commun. Stat.-Simul. C. **47**, 1382–1396 (2018)
86. Henderson, D.J., Parmeter, C.F.: Model averaging over nonparametric estimators. In: Essays in Honor of Aman Ullah. Advances in Econometrics, vol. 36, pp. 539–560. Emerald Group Publishing Limited, UK (2016)
87. Hinde, J., Demtrio, C.G.B.: Overdispersion: models and estimation. Comput. Stat. Data Anal. **27**, 151–170 (1998)
88. Hjort, N.L., Claeskens, G.: Frequentist model average estimators. J. Am. Stat. Assoc. **98**, 879–945 (2003)
89. Hjort, N.L., Claeskens, G.: Rejoinder to the Discussion of Hjort, N.L., Claeskens, G.: Frequentist model average estimators. J. Am. Stat. Assoc. **98**, 938–945 (2003)
90. Hjort, N.L., Claeskens, G.: Focused information criteria and model averaging for the Cox hazard regression model. J. Am. Stat. Assoc. **101**, 1449–1464 (2006)
91. Hobbs, N.T., Hilborn, R.: Alternatives to statistical hypothesis testing in ecology: a guide to self teaching. Ecol. Appl. **16**, 5–19 (2006)
92. Holbrook, A., Gillen, D.: Estimating prediction error for complex samples (2017). arXiv preprint: arXiv:1711.04877
93. Hong, C.Y.: Focussed model averaging in generalised linear models. Thesis, Doctor of Philosophy, University of Otago (2018)
94. Hoogerheide, L., Kleijn, R., Ravazzolo, F., Van Dijk, H.K., Verbeek, M.: Forecast accuracy and economic gains from Bayesian model averaging using time-varying weights. J. Forecast. **29**, 251–269 (2010)
95. Hurvich, C.M., Tsai, C.-L.: Regression and time series model selection in small samples. Biometrika **76**, 297–307 (1989)
96. Hurvich, C.M., Tsai, C.-L.: Model selection for extended quasi-likelihood models in small samples. Biometrics **51**, 1077–1084 (1995)

97. Ishiguro, M., Sakamoto, Y.: WIC: An Estimation-free Information Criterion. Research Memorandum, Institute of Statistical Mathematics, Tokyo (1991)
98. Ishiguro, M., Sakamoto, Y., Kitagawa, G.: Bootstrapping log likelihood and EIC, an extension of AIC. Ann. Inst. Stat. Math. **49**, 411–434 (1997)
99. Jacobs, R.A.: Methods for combining experts' probability assessments. Neural Comput. **7**, 867–888 (1995)
100. Jensen, S.M., Ritz, C.: Simultaneous inference for model averaging of derived parameters. Risk Anal. **35**, 68–76 (2015)
101. Jiang, J., Rao, J.S., Gu, Z., Nguyen, T.: Fence methods for mixed model selection. Ann. Stat. **36**, 1669–1692 (2008)
102. Jin, S., Ankargren, S.: Frequentist model averaging in structural equation modelling. Psychometrika (2018). https://doi.org/10.1007/s11336-018-9624-y
103. Johnson, W.O.: Discussion of Hjort, N.L., Claeskens, G.: Frequentist model average estimators. J. Am. Stat. Assoc. **98**, 919–921 (2003)
104. Jullum, M., Hjort, N.L.: Parametric or nonparametric: the FIC approach. Stat. Sin. **27**, 951–981 (2017)
105. Kabaila, P., Leeb, H.: On the large-sample minimal coverage probability of confidence intervals after model selection. J. Am. Stat. Assoc. **101**, 619–629 (2006)
106. Kabaila, P., Welsh, A.H., Abeysekera, W.: Model-averaged confidence intervals. Scand. J. Stat. **43**, 35–48 (2016)
107. Kabaila, P., Welsh, A.H., Mainzer, R.: The performance of model averaged tail area confidence intervals Commun. Stat-Theor. M. **46**, 10718–10732 (2016)
108. Kabaila, P., Wijethunga, C.: Confidence intervals centered on bootstrap smoothed estimators (2016). arXiv preprint: arXiv:1610.09802
109. Kabaila, P.: On the minimum coverage probability of model averaged tail area confidence intervals. Can. J. Stat. **46**, 279–297 (2018)
110. Kapetanios, G., Mitchell, J., Price, S., Fawcett, N.: Generalised density forecast combinations. J. Econ. **188**, 150–165 (2015)
111. LeBlanc, M., Tibshirani, R.: Combining estimates in regression and classification. J. Am. Stat. Assoc. **91**, 1641–1650 (1996)
112. Lee, H., Jogesh Babu, G., Rao, C.R.R.: A jackknife type approach to statistical model selection. J. Stat. Plan. Inference **142**, 301–311 (2012)
113. Leeb, H., Pötscher, B.M.: Model selection and inference: facts and fiction. Econ. Theory **21**, 21–59 (2005)
114. Leeb, H., Pötscher, B.M.: Can one estimate the conditional distribution of post-model-selection estimators? Ann. Stat. **34**, 2554–2591 (2006)
115. Lemke, C., Budka, M., Gabrys, B.: Metalearning: a survey of trends and technologies. Artif. Intell. Rev. **44**, 117–130 (2015)
116. Lenkoski, A., Eicher, T.S., Raftery, A.E.: Two-stage Bayesian model averaging in endogenous variable models. Econ. Rev. **33**, 122–151 (2014)
117. Leung, G., Barron, A.R.: Information theory and mixing least squares regressions. IEEE Trans. Inf. Theory **52**, 3396–3410 (2006)
118. Li, C., Li, Q., Racine, J.S., Zhang, D.: Optimal model averaging of varying-coefficient models. Stat. Sin. (2018). https://doi.org/10.5705/ss.202017.0034
119. Li, J., Xia, X., Wong, W.K., Nott, D.: Varying-coefficient semiparametric model averaging prediction. Biometrics (2018). https://doi.org/10.1111/biom.12904
120. Liang, H., Wu, H., Zou, G.: A note on conditional AIC for linear mixed-effects models. Biometrika **95**, 773–778 (2008)
121. Liang, H., Zou, G., Wan. A.T.K., Zhang, X.: Optimal weight choice for frequentist model average estimators: J. Am. Stat. Assoc. **106**, 1053–1066 (2011)
122. Lieb, L., Smeekes, S.: Inference for impulse responses under model uncertainty (2017). arXiv preprint: arXiv:1709.09583
123. Lin, B., Wang, Q., Zhang, J., Pang, Z.: Stable prediction in high-dimensional linear models. Stat. Comput. **27**, 1401–1412 (2017)

124. Link, W., Barker, R.: Model weights and the foundations of multimodel inference. Ecology **87**, 2626–2635 (2006)
125. Liu, Q., Okui, R.: Heteroscedasticity-robust C_p model averaging. Econ. J. **16**, 463–472 (2013)
126. Liu, S., Yang, Y.: Combining models in longitudinal data analysis. Ann. Inst. Stat. Math. **64**, 233–254 (2012)
127. Liu, S., Yang, Y.: Mixing partially linear regression models. Sankhyā **75**, 74–95 (2013)
128. Liu, C.A.: Distribution theory of the least squares averaging estimator. J. Econ. **186**, 142–159 (2015)
129. Liu, Q., Okui, R., Yoshimura, A.: Generalized least squares model averaging. Econ. Rev. **35**, 1692–1752 (2016)
130. Longford, N.T.: An alternative to model selection in ordinary regression. Stat. Comput. **13**, 67–80 (2003)
131. Longford, N.T.: An alternative analysis of variance. SORT Stat. Oper. Res. T. **32**, 77–92 (2008)
132. Lu, X., Su, L.: Jackknife model averaging for quantile regressions. J. Econ. **188**, 40–58 (2015)
133. Lumley, T., Scott, A.: AIC and BIC for modeling with complex survey data. J. Surv. Stat. Methodol. **3**, 1–18 (2015)
134. Lv, J., Liu, J.S.: Model selection principles in misspecified models. J. R. Stat. Soc. Ser. B (Stat. Methodol.) **76**, 141–167 (2014)
135. Magnus, J.R., Wan, A.T.K., Zhang, X: Weighted average least squares estimation with non-spherical disturbances and an application to the Hong Kong housing market. Comput. Stat. Data Anal. **55**, 1331–1341 (2011)
136. Magnus, J.R., De Luca, G.: Weighted-average least squares (WALS): a survey. J. Econ. Surv. **30**, 117–148 (2016)
137. Mallows, C.L.: Some comments on Cp. Technometrics **42**, 87–94 (2000)
138. Martins, L.F., Gabriel, V.J.: Linear instrumental variables model averaging estimation. Comput. Stat. Data. Anal. **71**, 709–724 (2014)
139. McQuarrie, A., Shumway, R., Tsai, C.-L.: The model selection criterion AICu. Stat. Probabil. Lett. **34**, 285–292 (1997)
140. McQuarrie, A.D.R., Tsai, C.-L.: Regression and Time Series Model Selection. World Scientific, Singapore (1998)
141. Mead, R.: The Design of Experiments: Statistical Principles for Practical Applications. Cambridge University Press, Cambridge (1988)
142. Mitra, P., Lian, H., Mitra, R., Liang, H., Xie, M.: A general framework for frequentist model averaging (2018). arXiv preprint: arXiv:1802.03511
143. Moody, J.E.: The effective number of parameters: an analysis of generalization and regularization in nonlinear learning systems. In: Moody, J.E., Hanson, S.J., Lippmann, R.P. (eds.) Advances in Neural Information Processing Systems, vol. 4, pp. 847–854. Morgan Kaufmann, San Mateo, California (1992)
144. Müller, S., Scealy, J.L., Welsh, A.H.: Model selection in linear mixed models. Stat. Sci. **28**, 135–167 (2013)
145. Murata, N., Yoshizawa, S., Amari, S.: Network information criterion-determining the number of hidden units for artificial neural network models. IEEE Trans. Neural Netw. **5**, 865–872 (1994)
146. Murray, K., Conner, M.M.: Methods to quantify variable importance: implications for the analysis of noisy ecological data. Ecology **90**, 348–355 (2009)
147. Naftaly, U., Intrator, N., Horn, D.: Optimal ensemble averaging of neural networks. Network-Comp. Neural **8**, 283–296 (1997)
148. Nakagawa, S., Freckleton, R.P.: Model averaging, missing data and multiple imputation: a case study for behavioural ecology. Behav. Ecol. Sociobiol. **65**, 103–116 (2011)
149. Neath, A.A., Cavanaugh, J.E., Weyhaupt, A.G.: Model evaluation, discrepancy function estimation, and social choice theory. Comput. Stat. **29**, 1–19 (2014)
150. Owen, A.B.: Small sample central confidence intervals for the mean. Technical Report 302, Department of Statistics, Stanford University (1988)

151. Polley, E.C., van der Laan, M.J.: Super learner in prediction. UC Berkeley Division of Biostatistics Working Paper Series. Working Paper 266 (2010). http://biostats.bepress.com/ucbbiostat/paper266

152. Poeter, E.P., Hill, M.C.: MMA, a computer code for multi-model analysis. U.S. Geological Survey Techniques and Methods TM6-E3. Reston, Virginia (2007)

153. Pötscher, B.M.: The distribution of model averaging estimators and an impossibility result regarding its estimation. Inst. Math. S. **52**, 113–129 (2006)

154. Quenouille, M.H.: Notes on bias in estimation. Biometrika **43**, 353–360 (1956)

155. R Core Team: R: A language and environment for statistical computing. R Foundation for Statistical Computing, Vienna, Austria (2017). https://www.R-project.org/

156. Raftery, A.E., Zheng, Y.: Discussion of Hjort, N.L., Claeskens, G.: Frequentist model average estimators. J. Am. Stat. Assoc. **98**, 931–938 (2003)

157. Rao, J.S., Tibshirani, R.: The out-of-bootstrap method for model averaging and selection. University of Toronto (1997)

158. Richards, S.A.: Testing ecological theory using the information-theoretic approach: examples and cautionary results. Ecology **86**, 2805–2814 (2005)

159. Ripley, B.D.: Selecting amongst large classes of models. In: Adams, N., Crowder, M., Hand, D.J., Stephens, D. (eds.) Methods and Models in Statistics: in Honor of Professor John Nelder, FRS, pp. 155–170. Imperial College Press, London (2004)

160. Rubin, D.B.: Inference and missing data. Biometrika **63**, 581–592 (1976)

161. Saefken, B., Kneib, T., van Waveren, C.-S., Greven, S.: A unifying approach to the estimation of the conditional Akaike information in generalized linear mixed models. Electron. J. Stat. **8**, 201–225 (2014)

162. Sapp, S., van der Laan, M.J., Canny, J.: Subsemble: an ensemble method for combining subset-specific algorithm fits. J. Appl. Stat. **41**, 1247–1259 (2014)

163. Schapire, R.E.: The strength of weak learnability. Mach. Learn. **5**, 197–227 (1990)

164. Schomaker, M.: Shrinkage averaging estimation. Stat. Pap. **53**, 1015–1034 (2012)

165. Schomaker, M., Wan, A.T.K., Heumannm, C.: Frequentist model averaging with missing observations. Comput. Stat. Data. Anal. **54**, 3336–3347 (2010)

166. Schomaker, M., Heumannm, C.: Model selection and model averaging after multiple imputation. Comput. Stat. Data. Anal. **71**, 758–770 (2014)

167. Schomaker, M., Heumann, C.: When and when not to use optimal model averaging (2018). arXiv preprint: arXiv:1802.04589

168. Shan, K., Yang, Y.: Combining regression quantile estimators. Stat. Sin. **19**, 1171–1191 (2009)

169. Shang, J., Cavanaugh, J.E.: Bootstrap variants of the Akaike information criterion for mixed model selection. Comput. Stat. Data. Anal. **52**, 2004–2021 (2008)

170. Shen, X., Huang, H.-C., Ye, J.: Adaptive model selection and assessment for exponential family distributions. Technometrics **46**, 306–317 (2004)

171. Shen, X., Huang, Huang.-C.: Optimal model assessment, selection, and combination. J. Am. Stat. Assoc. **101**, 554–568 (2006)

172. Shibata, R.: Bootstrap estimate of Kullback-Leibler information for model selection. Stat. Sin. **7**, 375–394 (1997)

173. Smith, A.C., Koper, N., Francis, C.M., Fahrig, L.: Confronting collinearity: comparing methods for disentangling the effects of habitat loss and fragmentation. Landscape Ecol. **24**, 1271–1285 (2009)

174. Smyth, P., Wolpert, D.: Linearly combining density estimators via stacking. Mach. Learn. **36**, 59–83 (1999)

175. Stone, M.: Cross-validatory choice and assessment of statistical predictions. J. R. Stat. Soc. Ser. B. (Methodol.) **36**, 111–147 (1974)

176. Stone, M.: An asymptotic equivalence of choice of model by cross-validation and Akaike's criterion. J. R. Stat. Soc. Ser. B. (Methodol.) **39**, 44–47 (1977)

177. Sugiura, N.: Further analysts of the data by Akaike's information criterion and the finite corrections: further analysts of the data by Akaike's. Commun. Stat. Theory **7**, 13–26 (1978)

178. Takeuchi, K.: Distribution of informational statistics and a criterion of model fitting. Suri-Kagaku (Math. Sci.) **153**, 12–18 (1976)
179. Timmermann, A.: Forecast combinations. In: Elliott, G., Granger, C.W.J., Timmermann, A. (eds.) Handbook of Economic Forecasting, pp. 135–196. Elsevier, Amsterdam (2006)
180. Ting, K.M., Witten, I.H.: Issues in stacked generalization. J. Artif. Intell. Res. **10**, 271–289 (1999)
181. Turek, D., Fletcher, D.: Model-averaged Wald confidence intervals. Comput. Stat. Data. Anal. **56**, 2809–2815 (2012)
182. Turek, D.: Comparison of the frequentist MATA confidence interval with Bayesian model-averaged confidence intervals. J. Probab. Stat. (2015). https://doi.org/10.1155/2015/420483
183. Ullah, A., Wang, H.: Parametric and nonparametric frequentist model selection and model averaging. Econ. J. **1**, 157–179 (2013)
184. Vaida, F., Blanchard, S.: Conditional Akaike information for mixed-effects models. Biometrika **92**, 351–370 (2005)
185. van der Laan, M.J., Dudoit, S., Keles, S.: Asymptotic optimality of likelihood-based cross-validation. Stat. Appl. Genet. Mol. **3**, Article 4 (2004)
186. van der Laan, M.J., Polley, E.C., Hubbard, A.E.: Super learner. Stat. Appl. Genet. Mol. Biol. **6**, 1–23 (2007)
187. Wagenmakers, E.-J., Farrell, S., Ratcliff, R.: Letter to the editors. Biometrics **60**, 281–283 (2004)
188. Wager, S., Hastie, T., Efron, B.: Confidence intervals for random forests: the jackknife and the infinitesimal jackknife. J. Mach. Learn. Res. **15**, 1625–1651 (2014)
189. Wallis, K.F.: Combining density and interval forecasts: a modest proposal. Oxford B. Econ. Stat. **67**, 983–994 (2005)
190. Wan, A.T.K., Zhang, X., Zou, G.: Least squares model averaging by Mallows criterion. J. Econ. **156**, 277–283 (2010)
191. Wan, A.T.K., Zhang, X., Wang, S.: Frequentist model averaging for multinomial and ordered logit models. In. J. Forecast. **30**, 118–128 (2014)
192. Wang, H., Zou, G., Wan, A.T.K.: Model averaging for varying-coefficient partially linear measurement error models. Electron. J. Stat. **6**, 1017–1039 (2012)
193. Wang, H., Zhou, S.Z.F.: Interval estimation by frequentist model averaging. Commun. Stat. Theory **42**, 4342–4356 (2013)
194. Wang, H.Y., Chen, X., Flournoy, N.: The focused information criterion for varying-coefficient partially linear measurement error models. Stat. Pap. 1–15. Springer, Heidelberg (2014)
195. Wang, H., Li, Y., Sun, J.: Focused and model average estimation for regression analysis of panel count data. Scand. J. Stat. **42**, 732–745 (2015)
196. Wedderburn, R.W.M.: Quasi-likelihood functions, generalized linear models, and the Gauss-Newton method. Biometrika **61**, 439–447 (1974)
197. White, H.: Maximum likelihood estimation of misspecified models. Econometica **50**, 1–25 (1982)
198. Wolpert, D.H.: Stacked generalization. Neural Netw. **5**, 241–259 (1992)
199. Wood, S.N.: Core Statistics. Cambridge University Press, Cambridge (2015)
200. Xie, T.: Prediction model averaging estimator. Econ. Lett. **131**, 5–8 (2015)
201. Xu, R., Gamst, A., Donohue, M., Vaida, F., Harrington, D.P.: Using profile likelihood for semiparametric model selection with application to proportional hazards mixed models. Harvard University Biostatistics Working Paper Series, Paper 43 (2006). http://biostats.bepress.com/harvardbiostat/paper43/
202. Xu, G., Wang, S., Huang, J.Z.: Focused information criterion and model averaging based on weighted composite quantile regression. Scand. J. Stat. **41**, 365–381 (2014)
203. Xu, R., Mehrotra, D.V., Shaw, P.A.: Incorporating baseline measurements into the analysis of crossover trials with timetoevent endpoints. Stat. Med. (2018). https://doi.org/10.1002/sim.7834
204. Yang, Y.: Adaptive regression by mixing. J. Am. Stat. Assoc. **96**, 574–588 (2001)

205. Yang, Y.: Regression with multiple candidate models: selecting or mixing? Stat. Sin. **13**, 783–809 (2003)

206. Yang, Y.: Can the strengths of AIC and BIC be shared? A conflict between model indentification and regression estimation. Biometrika **92**, 937–950 (2005)

207. Ye, J.: On measuring and correcting the effects of data mining and model selection. J. Am. Stat. Assoc. **93**, 120–131 (1998)

208. Yu, D., Yau, K.K.W.: Conditional Akaike information criterion for generalized linear mixed models. Comput. Stat. Data. Anal. **56**, 629–644 (2012)

209. Yu, Y., Thurston, S.W., Hauser, R., Liang, H.: Model averaging procedure for partially linear single-index models. J. Stat. Plan. Infer. **143**, 2160–2170 (2013)

210. Yu, W., Xu, W., Zhu, L.: Transformation-based model averaged tail area inference. Comput. Stat. **29**, 1713–1726 (2014)

211. Yu, D., Zhang, X., Yau, K.K.W.: Asymptotic properties and information criteria for misspecified generalized linear mixed models. J. R. Stat. Soc. Ser. B (Methodol.) (2018). https://doi.org/10.1111/rssb.12270

212. Yuan, Z., Yang, Y.: Combining linear regression models. J. Am. Stat. Assoc. **100**, 1202–1214 (2005)

213. Yuan, Z., Ghosh, D.: Combining multiple biomarker models in logistic regression. Biometrics **64**, 431–439 (2008)

214. Zeng, J.: Model-Averaged Confidence Intervals. (Thesis, Doctor of Philosophy). University of Otago (2013)

215. Zeng, J., Cheng, W., Hu, G., Ronga, Y.: Model averaging procedure for varying-coefficient partially linear models with missing responses. J. Korean Stat. Soc. **47**, 379–394 (2018)

216. Zhang, X., Liang, H.: Focused information criterion and model averaging for generalized additive partial linear models. Ann. Stat. **39**, 174–200 (2011)

217. Zhang, C., Ma, Y.: (eds.) Ensemble Machine Learning: Methods and Applications. Springer, New York (2012)

218. Zhang, X., Wan, A.T.K., Zhou, S.Z.: Focused information criteria, model selection, and model averaging in a Tobit model with a nonzero threshold. J. Bus. Econ. Stat. **30**, 132–142 (2012)

219. Zhang, X., Wan, A.T.K., Zou, G.: Model averaging by jackknife criterion in models with dependent data. J. Econ. **174**, 82–94 (2013)

220. Zhang, X., Zou, G., Carroll, R.J.: Model averaging based on Kullback-Leibler distance. Stat. Sin. **25**, 1583–1598 (2015)

221. Zhang, X.: Consistency of model averaging estimators. Econ. Lett. **130**, 120–123 (2015)

222. Zhang, Y., Yang, Y.: Cross-validation for selecting a model selection procedure. J. Econ. **187**, 95–112 (2015)

223. Zhang, X., Yu, D., Zou, G., Liang, H.: Optimal model averaging estimation for generalized linear models and generalized linear mixed-effects models. J. Am. Stat. Assoc. **111**, 1775–1790 (2016)

224. Zhang, Q., Duan, X., Ma, S.: Focused information criterion and model averaging with generalized rank regression. Stat. Probabil. Lett. **122**, 11–19 (2017)

225. Zhao, N., Zhao, Z., Liao, S.: Probabilistic model combination for support vector machine using positive-definite kernel-based regularization path. In: Wang, Y., Li, T. (eds.) Foundations of Intelligent Systems. Advances in Intelligent and Soft Computing, vol. 122, pp. 201–206. Springer, Heidelberg (2011)

226. Zhao, S., Zhang, X., Gao, Y.: Model averaging with averaging covariance matrix. Econ. Lett. **145**, 214–217 (2016)

227. Zhao, S., Ullah, A., Zhang, X.: A class of model averaging estimators. Econ. Lett. **162**, 101–106 (2018)

228. Zou, G., Wan, A.T.K., Wu, X., Chen, T.: Estimation of regression coefficients of interest when other regression coefficients are of no interest: the case of non-normal errors. Stat. Probabil. Lett. **77**, 803–810 (2007)

Chapter 4
Summary and Future Directions

Abstract We provide an overview of the key ideas and results in Bayesian and frequentist model averaging, and suggest directions for future research.

4.1 Summary of Key Points

Estimation not identification
Model averaging is an estimation tool, and identification of a true model is therefore not directly relevant. In many settings, an assessment of the sensitivity of the estimates to the choice of model will also be useful.

Parameter of interest
This should have the same interpretation in all models, and averaging of regression coefficients is therefore unlikely to be relevant.

Model-redundancy
If we use a uniform model-prior in classical BMA, model-redundancy can lead to dilution of some of the prior model probabilities. This problem does not arise in prediction-based BMA as it does not require a model-prior. Likewise, in FMA model-redundancy can cause AIC weights to be diluted. The simplex-constraint used in AIC(w) and stacking alleviates this problem.

Interval estimation
No interval currently guarantees good coverage unless it is equivalent to the interval from the largest model. The MATA interval has the advantage that it can be based on any method for calculating a single-model confidence interval.

Choice of scale
BMA is transformation-invariant whereas FMA is not. However, there is usually a natural scale on which to perform FMA, such as the linear-predictor scale in a GLM.

© The Author(s), under exclusive licence to Springer-Verlag GmbH, DE, part of Springer Nature 2018

D. Fletcher, *Model Averaging*, SpringerBriefs in Statistics, https://doi.org/10.1007/978-3-662-58541-2_4

Summing model weights

Summing model weights does not provide a useful measure of the importance of a predictor; comparison of model-averaged estimates for specific values of the predictors is preferable. A similar comment applies to posterior inclusion-probabilities in BMA.

Mixed models

Mixed-model versions of AIC(w) and stacking are possible, as are hierarchical-versions of WAIC and BSP.

Choice of model set

Using a set of singleton models, each involving a single predictor, appears to be a promising approach when there are many predictors.

Choice of Bayesian method

Use of WAIC or BSP weights is preferable to classical BMA, as the focus is then on prediction rather than identification of the true model. Use of these weights also avoids problems with the calculation of posterior model probabilities and the sensitivity of these probabilities to priors on the parameters.

Choice of frequentist method

If computational effort is not an issue, stacking is a good choice. AIC(w) is a good alternative for large n, but is less robust. Both methods also have a nice interpretation in terms of a meta-model.

4.2 Future Directions

Confidence intervals

Work is needed on the best approach to calculating a model-averaged confidence interval. This might involve weights that are optimal with respect to interval coverage and width, as suggested by [2] in the Bayesian setting. Work is also needed on assessing when simply using the confidence interval from the largest model is best.

Confidence distributions

A clear advantage of BMA is use of a posterior distribution to summarise the results. In FMA it would be useful to have a model-averaged version of a confidence distribution, which provides a summary of all possible confidence intervals [1].

Alternative versions of WAIC and BSP

A weighted version of WAIC, analogous to AIC(w) in FMA, would be useful for the same reason that AIC(w) weights seem preferable to those based on AIC. Likewise, for GLMs, a version of BSP that involves weighting the linear predictor for each model, analogous to stacking in FMA, might be useful.

Comparison with shrinkage

It would be good to have simulation studies that compare model averaging methods, such as AIC(w) and stacking, with shrinkage methods, such as the lasso.

References

1. Xie, M.G., Singh, K.: Confidence distribution, the frequentist distribution estimator of a parameter: a review. Int. Stat. Rev. **81**, 3–39 (2013)
2. Yao, Y., Vehtari, A., Simpson, D., Gelman, A.: Using stacking to average Bayesian predictive distributions. Bayesian Anal. (2018). https://doi.org/10.1214/17-BA1091

Index

© The Author(s), under exclusive licence to Springer-Verlag GmbH, DE, part of Springer Nature 2018

D. Fletcher, *Model Averaging*, SpringerBriefs in Statistics, https://doi.org/10.1007/978-3-662-58541-2

MIX
Papier aus verantwortungsvollen Quellen
Paper from responsible sources
FSC® C105338

If you have any concerns about our products,
you can contact us on
ProductSafety@springernature.com

In case Publisher is established outside the EU,
the EU authorized representative is:
Springer Nature Customer Service Center GmbH
Europaplatz 3, 69115 Heidelberg, Germany

Printed by Libri Plureos GmbH
in Hamburg, Germany